Emergency Information

Name of Chemist _____

Address (optional) _____

Telephone Number (optional) _____

Laboratory Information

Course Number _____

Lab Section Number _____

Instructor's Name _____

Laboratory Building and Room Number _____

Desk Number _____

Stock Room Number _____

Location of Safety Equipment Nearest to Your Laboratory Bench

Main Power Shut-Off Switch _____

Main Gas Shut-Off Valve _____

Telephone _____

First Aid Kit _____

Fire Extinguisher _____

Fire Blanket _____

Safety Shower _____

Eye Wash Fountain _____

Campus Medical Office _____

Local Emergency Telephone Numbers

Campus Safety _____

Ambulance _____

Fire Department _____

Police _____

Laboratory Manual

Third Edition

Charles D. Anderson Pacific Lutheran University

David B. Macaulay William Rainey Harper College

Molly M. Bloomfield Oregon State University

Joseph M. Bauer William Rainey Harper College

To Accompany

General, Organic, and Biological Chemistry: An Integrated Approach

Fourth Edition

Kenneth W. Raymond

ISBN 978-1-118-42426-1

Printed in the United States of America

Printed and bound by Strategic Content Imaging

A Note from the Publisher

Laboratory Manual to accompany General, Organic, and Biological Chemistry: An Integrated Approach, highlights the relevance and application of basic chemical principles to biological systems. The laboratory experiments are designed to relate to the student's personal experience and use many common household and commercial products.

This manual accompanies *General, Organic, and Biological Chemistry: An Integrated Approach*, Fourth Edition, by Kenneth W. Raymond. The laboratory exercises are closely integrated with the text subject matter, with specific text references available for the majority of the experiments. The manual assumes no prior student experience in the chemistry laboratory.

ORGANIZATION OF THE MANUAL

Safety in the Laboratory

This section provides safety rules and precautions. It contains the discussion on MSDS sheets and NFPA hazard ratings.

Fundamental Laboratory Operations

This highly illustrated section demonstrates and explains the most commonly performed laboratory operations. Frequent reference is made to various portions of this section throughout the manual.

Twenty Laboratory Experiments and an Exercise on Chemical Periodicity

These experiments supply the instructor with a wide selection for use during a one-semester course, while also providing experiments to cover the laboratory portion of a two-semester course. Most experiments can be completed within two hours; by expanding pre-laboratory and post-laboratory discussions, a three-hour period can be utilized. Each experiment consists of the following parts: an Introduction, Procedure, Prelab Questions, a Report, and Related Questions. The **Introduction** explains the chemical principles relevant to the experiment; in many instances examples pertinent to the allied heath fields are provided. Most experiments make a reference to specific chapters in the 4th edition of Kenneth Raymond's *General, Organic, and Biological Chemistry: An Integrated Approach*. The Procedure section presents the instructions for each experiment. **Prelab Questions** are to be completed before the start of the laboratory period. Experimental data, calculations, and conclusions are to be recorded in the **Report** section, which also includes specific questions about the student's observations. Additional questions that require the student to apply some of the experimental principles to other problems are found in the **Related Questions** section of each experiment. The answers to these questions may require reference to the text or other sources such as *The Merck Index* or the *CRC Handbook of Chemistry and Physics*.

INSTRUCTOR'S MANUAL

The Instructor's Manual provides teaching hints and suggestions, expected data and results, answers to questions, a list of supplies and chemicals for each experiment, and directions for preparing solutions.

The Instructor's Manual is available for download on the text website: www.wiley.com/college, or by request through your Wiley Sales Representative.

ACKNOWLEDGEMENTS

We would like to thank Kenneth Raymond for his helpful feedback during the development of this project, and to acknowledge Jennifer Yee and Barbara Russiello for their excellent help and concerted efforts in producing this new edition of the manual, and Bette Kreuz for her expert analysis and careful review of these experiments.

Table of Contents

Safety in the Laboratory

Be certain to read these instructions carefully **before attending the first laboratory session**.

BEFORE COMING TO THE LABORATORY

Before you come into the laboratory, carefully read and study the laboratory experiment to be performed. The most common reason that students do not complete an experiment in the time allowed is lack of prelaboratory preparation. Your instructor will provide additional instructions for an experiment when needed, and will let you know whether the experiment is to be performed individually, in pairs, or in groups. You may be requested to complete and hand in the Prelab Questions before performing an experiment. Record all data for each experiment on the Report as soon as you have completed a particular laboratory operation. The Related Questions may be completed after the laboratory period.

RULES FOR WORKING IN THE LABORATORY

At the Laboratory Station

1. Turn off hotplates (including the magnetic stirring motor, if present) or any other electrical apparatus when not in use. Unplug the apparatus when you are finished with the experiment.
2. When heating beakers or flasks with a Bunsen burner, use a wire gauze under the container for more even distribution of heat and more rapid heating. Turn of Bunsen burner when not in use.
3. If any hazardous reagents are spilled, notify your instructor at once.

Handling Chemicals

1. Check odors cautiously; never taste a chemical.
2. To avoid spattering of acids which can cause burns, always add acid to water—never add water to acid.
3. Before obtaining any reagents, carefully read the labels on the bottles twice. Many chemicals have similar names.
4. Do not remove a reagent from the designated dispensing area. Use the appropriate container (test tube, beaker, etc.) for obtaining chemicals.
5. To avoid unnecessary waste, obtain only the amount of chemicals called for in an experiment. Your instructor will tell you the proper procedure for dispensing liquids and solids. See Fundamental Laboratory Operations, Part 1 on page 8.
6. Never return unused chemicals to the original dispensing bottle.
7. Follow the instructor's directions for disposal of chemicals and any waste produced in the experiment. If no specific directions are given, check with the instructor.

SAFETY PRECAUTIONS

1. Perform only the experiment assigned; do not experiment on your own.
2. Carefully note any precautions stated in the laboratory experiment, especially those written in capital letters.
3. Always wear safety goggles of the specific type required in your laboratory. Do not wear sandals or open-toed shoes. Do not wear clothes with loose, long sleeves.
4. Since many reagents are flammable or toxic, smoking, eating, chewing gum, or drinking are not permitted in the laboratory.
5. Hair can easily catch on fire; if you have long hair, tie it back.
6. Know the location of the fire extinguisher, fire blanket, eye wash, and safety shower. Your instructor will describe their use.
7. Contact lenses can pose a specific hazard in the chemistry laboratory. Check with your instructor as to the policy in your laboratory regarding contact lenses. Contact lenses may not be allowed. If you wear contact lenses, be especially wary of splashing a chemical in your eyes. It is possible for liquid reagents to seep under the lens so that flushing your eyes will not remove all the chemical. The lens should be removed as quickly as possible so that both the lens and eye can be thoroughly washed. Plastic contact lenses can also absorb vapors of organic solvents.
8. When noxious vapors are likely to be produced or when volatile or flammable chemicals are used, perform the laboratory operations under a fume hood.
9. Be especially conscious of hot objects and broken glassware. Burns and cuts are the most common types of injury in the laboratory. Promptly report any injury, no matter how slight, to your instructor. Also report broken or chipped glassware.
10. When using a burner to heat a liquid in a test tube, clamp the tube at the top, and concentrate the flame near the surface of the liquid. Do not point the open end of the test tube toward another person. Do not heat graduated cylinders or bottles or apply an open flame to a watch glass. See Fundamental Laboratory Operations, Part 6, page 18.
11. When heating liquids in other containers, be sure to add either a boiling stone (black SiC if acidic) or magnetic stir bar to the liquid to avoid bumping. Never add a boiling stone or stir bar to a hot liquid as this may cause dangerous spattering of the liquid out of the container. Never heat a glass container to dryness. See Fundamental Laboratory Operations, page 19.
12. When using hotplates or any electrical apparatus, avoid contact between water and the apparatus or any of the electrical connections.
13. Be sure to use the proper devices for holding hot beakers, test tubes, crucibles, etc. Your instructor will demonstrate the correct procedure.
14. If you do not understand the instructions in this manual, ask your instructor for clarification before performing the procedure.

LABELING OF CHEMICALS

Every chemical in a chemistry laboratory must be properly labeled. Many chemicals have similar names and you should read the name twice. If a chemical is a solution, the concentration will also appear on the label. Solution concentration is commonly described by molarity (e.g., 6 M HCl) or by percent concentration (e.g., 0.9% NaCl).

Materials are often labeled with information about the hazards of the chemicals. One of the most common systems was developed by the **National Fire Protection Association* (NFPA)**. NFPA labels describe four hazards—health, fire, reactivity, and specific hazards. For each hazard a numerical rating of 0 to 4 is given. Each hazard rating is displayed on a color-coded diamond-shaped label, divided into 4 smaller colored labels. See Figure 0-1.

---

*See *Fire Protection Guide to Hazardous Materials*, 2010 edition; G.R. Colonna, Ed.; National Fire Protection Association, Quincy, MA.

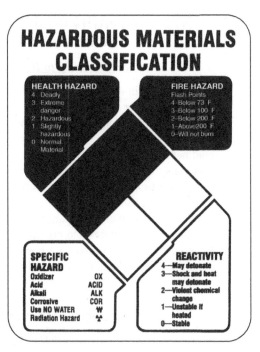

Figure 0-1 NFPA hazardous materials classification system.

NFPA Health Hazard Ratings
Color Code: BLUE

Rating	*Type of Possible Injury*
4	Materials that could cause death or major residual injury.
3	Materials that on short exposure could cause serious temporary or residual injury.
2	Materials that on intense or continued but not chronic exposure could cause temporary incapacitation or possible residual injury.
1	Materials that on exposure could cause irritation but only minor residual injury.
0	Materials that under fire conditions would offer no hazard beyond that of ordinary combustible material.

NFPA Flammability Hazard Ratings
Color Code: RED

Rating	*Susceptibility of Materials to Burning*
4	Materials that will rapidly or completely vaporize at atmospheric pressure and normal ambient temperature, or that are readily dispersed in air and that burn readily.
3	Liquids and solids that can be ignited under almost all ambient temperature conditions.
2	Materials that must be moderately heated or exposed to relatively high ambient temperature before ignition can occur.
1	Materials that must be preheated before ignition can occur.
0	Materials that will not burn.

NFPA Reactivity (Stability) Hazard Ratings
Color Code: YELLOW

Rating	Susceptibility to Release of Energy
4	Materials that in themselves are readily capable of detonation or of explosive decomposition or reaction at normal temperatures and pressures.
3	Materials that in themselves are capable of detonation or explosion or reaction but which require a strong initiating source or which must be heated under confinement before initiation or which react explosively with water.
2	Materials that readily undergo violent chemical change at elevated temperatures and pressures or which react violently with water or which may form explosive mixtures with water.
1	Materials that in themselves are normally stable, but which can become unstable at elevated temperatures and pressures.
0	Materials that in themselves are normally stable, even under fire exposure conditions, and which are not reactive with water.

NFPA Specific Hazards

Symbol	Special Hazard
W̶	Materials that demonstrate unusual reactivity with water.
OX	Materials that demonstrate oxidizing properties.

The NFPA system was developed for the needs of firefighters, not for chemists. The NFPA ratings represent hazard under fire conditions as opposed to normal laboratory use, a situation that may distort hazard characteristics. Many chemicals are not rated. Several chemical companies have also developed similar, more comprehensive, rating systems but thus far they are not as widely used as the NFPA system, nor do they use standardized colors and ratings.

MATERIAL SAFETY DATA SHEETS

Material Safety Data Sheets (MSDSs) are safety reports that must be supplied by the manufacturer with each chemical purchased. An MSDS lists 16 pertinent points that enumerate all the known vital information regarding the safe shipping, storage, use, handling, and disposing of a chemical. MSDS sheets, while providing specific information and first aid measures, also provide a wealth of information about their physical and chemical properties. Any institution must have MSDSs readily available for all the chemicals it uses.

Because of varying safety considerations in different countries, the United Nations has lead in the development of a new, eventually universal form of safety data sheet. These SDSs are similar to MSDSs that are in use in the United States but include additional features that are gradually being incorporated in the data sheets prepared by US companies. Included are nine pictograms that indicate hazardous aspects of a chemical and that are beginning to be used on bottle labels. Further information on the new Globally Harmonious System (GHS) are conveniently available on the Sigma Aldrich website:

http://www.sigmaaldrich.com/safety-center/globally-harmonized.html#links *

---

*Wiley and the authors have made every effort to ensure that these links are actively maintained. Occasionally, some sites may be taken offline by their owners without our knowledge.

Another useful source of general and hazard information is *The Merck Index*, which is published (in print and on-line form) by the pharmaceutical company of that name. It contains entries on about 40,000 substances listed alphabetically by a well known name. Virtually all synonyms are given in the Name Index in the back, making this reference especially easy to use. Information provided on these substances includes formulas, structures, physical and physiological data, and uses, medical and non-medical. Its publication is now being assumed by the UK's Royal Society of Chemistry.

WASTE DISPOSAL

The Resource Conservation and Recovery Act (RCRA) gives the U.S. Environmental Protection Agency (EPA) the responsibility to regulate chemical waste. In the laboratory, we must do our part to comply with all federal, state, and local regulations affecting disposal of wastes. Your laboratory instructor will discuss the procedures for waste collection. Specific waste containers may be provided. Never dispose of anything in the sink or wastebasket unless specifically directed to do so.

This manual minimizes the production of wastes by using smaller quantities of chemicals whenever possible.

SUGGESTED EQUIPMENT (Revise to Meet Local Needs.)

Equipment	Quantity	Check-In	Check-Out
Apron	1		
Beaker, 150-mL	2		
Beaker, 250-mL	2		
Beaker, 400-mL	2		
Bottle, washing, plastic	1		
Brush, test-tube, large	1		
Brush, test-tube, small	1		
Cylinder, graduated, 10-mL	1		
Cylinder, graduated, 50-mL	1		
Crucible and lid, 15-mL	1		
Crucible tongs	1		
Dropper, medicine	4		
Evaporating dish, porcelain, 80-mm	2		
Flask, Erlenmeyer, 125-mL	2		
Flask, Erlenmeyer, 250-mL	2		
Forceps	1		
Funnel, 75-mm	1		
Reaction plate, 24-well	1		
Stirring rod, 200-mm	2		
Test-tube clamp	1		
Test-tube support (rack), wood or plastic	1		
Test tube, 10 × 75 mm	6		
Test tube, 16 × 150 mm	8		
Test tube, 25 × 150 mm	1		
Triangle	1		
Watch glass, 40-mm	1		
Watch-glass, 75-mm	2		
Wire gauze	1		

Nonreturnables

Filter paper, qualitative, 12.5-cm	1 pkg		
Litmus paper vial, blue	1		
Litmus paper vial, red	1		
Matches	1 box		

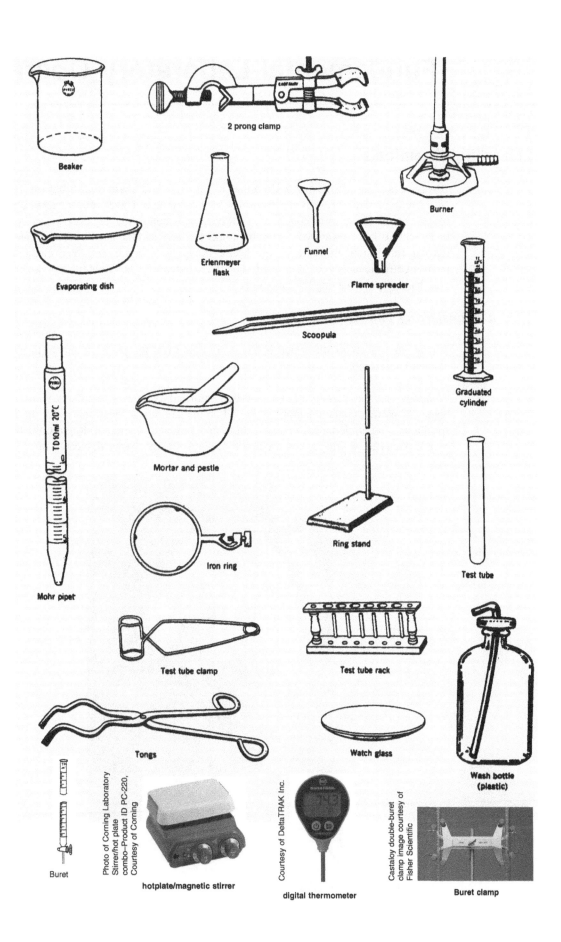

Beaker

2 prong clamp

Burner

Evaporating dish

Erlenmeyer flask

Funnel

Flame spreader

Scoopula

Graduated cylinder

Mohr pipet

Mortar and pestle

Iron ring

Ring stand

Test tube

Test tube clamp

Test tube rack

Tongs

Watch glass

Wash bottle (plastic)

Buret

Photo of Corning Laboratory Stirrer/hot plate combo–Product ID PC-220, Courtesy of Corning

hotplate/magnetic stirrer

Courtesy of DeltaTRAK Inc.

digital thermometer

Castaloy double-buret clamp image courtesy of Fisher Scientific

Buret clamp

Fundamental Laboratory Operations

1. Handling Chemicals

Always read the MSDS for a chemical to be used and then read label on its bottle twice.
Never remove the reagent bottle from the dispensing area. Never return excess chemicals to the
reagent bottle. Never taste, touch, or smell chemicals unless specifically directed to do so. Hold a
bottle with its label toward your palm to protect the label and the hand of the next user in case some
reagent runs down the side of the bottle. Keep your hands and face clean. If a chemical
inadvertently contacts your skin, wash the affected area with large amounts of water. Always wash
your hands before leaving the laboratory.

a. Transferring Powders and Crystals

Tilt the jar slightly as illustrated in Figure 0-2, and roll the jar back and forth until the desired
amount is obtained.

Withdraw only the quantity of chemical that you need, and never return any unused reagent to the
bottle. To prevent contamination, lay the lid on the laboratory bench inner-side up. Do not insert a
spatula into the bottle unless you are told to do so by your instructor. Replace the lid on the bottle
even if others are waiting to use the bottle. If you spill any chemical, clean it up at once.

Figure 0-2 Dispensing a solid from a reagent bottle by rolling back and forth.

b. Transferring Liquids

If the bottle has a screw cap, simply unscrew the cap and lay it on the laboratory bench inner-side up. If the bottle has a ground glass stopper as in Figure 0-3, loosen the stopper by gently twisting it. Some ground glass stoppers have a vertical ridge which should be used to hold the stopper between two fingers as in Figure 0-4. Other ground glass stoppers have a flat top—these should lay on the laboratory bench inner-side up. Never allow the inner side of the stopper to touch the bench; impurities may be picked up which can contaminate the liquid when the stopper is returned. Replace the stopper on the bottle even if others are waiting to use the bottle.

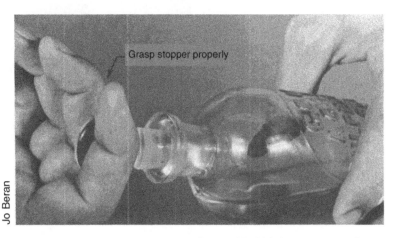

Figure 0-3 Remove the stopper by twisting and hold it between your fingers.

To transfer the liquid to another container, hold a stirring rod against the lip of the pouring bottle and pour down the stirring rod. The stirring rod should touch the wall of the receiving beaker as in Figure 0-4.

Figure 0-4 Use a stirring rod to transfer a liquid.

Figure 0-5 The stirring rod should touch the tip of the pouring beaker and the wall of the receiving beaker.

When transferring from a beaker, hold the stirring rod against the lip of the pouring beaker while touching the tip of the stirring rod to the side of the receiving beaker. See Figure 0-5.

2. Cleaning Glassware

Glassware must be clean to prevent errors caused by contaminants. Wash the glassware with a small amount of detergent and tap water. Use a brush to scrub the glassware. Rinse the glassware several times with tap water, then rinse once or twice with a small amount of deionized* water. While rinsing with deionized water, tip and roll the glassware and allow the deionized water to wet the entire inner surface of the glassware. If rinsing a pipet, buret, or other glassware with a tip, discard the rinse water through the tip. If the glassware is truly clean, the final rinse water will flow smoothly down the glass without leaving any drops clinging to the glass.

Invert the clean glassware on a paper towel to dry. Use a test tube rack or other support to prevent test tubes or flasks from tipping over. If time does not permit air drying, a paper towel may be used to dry the glassware—this can cause contamination and ought to be avoided if possible. Do not use compressed air to dry glassware since it is not very effective and is often contaminated with oil and water. Glassware can be placed in a drying oven to dry. Use insulated gloves in removing the glassware as it will be hot.

3. Use of Volumetric Glassware

Before use, all glassware should be thoroughly cleaned. See Fundamental Operation 2, above.

---

*Deionized water is water that has been treated to remove ions (charged particles) and other impurities. Deionized water is very similar to distilled water.

Graduated cylinders, pipets, burets, volumetric flasks, and syringes are the most commonly used glassware for measuring accurate volumes of liquids. Such glassware is available in an assortment of sizes, and each is used for a specific kind of laboratory operation. Beakers and flask are containers rather than measuring devices. The graduations on the sides of beakers and flasks provide only a rough estimate of volume and are not used for precise volume measurements.

All types of volumetric glassware have a slender, cylindrical shape in the measuring region, which causes the surface of most liquids to be curved downward. Take readings from the bottom of the curved surface (called the **meniscus**) with your eye at the same level. If the volumetric glassware that you are using has a number of graduations, estimate the volume as accurately as you can by noting the position of the meniscus between the graduations. See Figure 0-6.

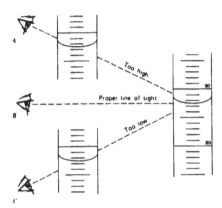

Figure 0-6 The proper method of reading the meniscus. The reading is 87.5 mL.

a. Graduated Cylinders

Graduated cylinders range in size from 1 mL to more than 1 L and provide a moderate degree of accuracy. They are the quickest and most commonly used of all volumetric glassware (see Figure 0-7). They are similar to kitchen measuring cups but their slender shape makes them more accurate. Before use, clean the cylinder as directed above.

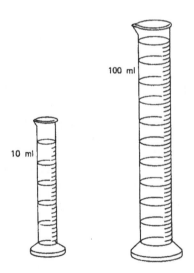

Figure 0-7 Graduated cylinders of different capacity.

b. Pipets

Pipets are useful for delivery of liquids when a greater degree of accuracy is required than is provided by a graduated cylinder. The two most frequently used pipets are the volumetric pipet and the serological pipet. See Figure 0-8. A volumetric pipet delivers only a single specific volume of liquid (for example 5 mL or 25 mL) and, therefore, has only one graduation mark. A serological pipet, on the other hand, has many graduation marks on its sides and may be used to deliver variable amounts of liquid. A variant is the Mohr pipet, which is not graduated to the tip. With it, volumes have to be measured between graduation marks at top and bottom.

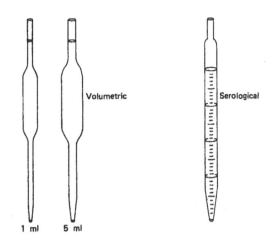

Figure 0-8 Volumetric and serological pipets.

Pipets marked "TC" (to contain) will contain the indicated volume when filled to the mark. Pipets marked "TD" (to deliver) will deliver the amount indicated on the pipets when the liquid flows out. A 10-mL TD pipet contains more than 10 mL—it delivers 10 mL and a small amount of liquid remains in the pipet. Volumetric pipets are usually marked "TD" and serological pipets are usually "TC."

In use, the liquid to be dispensed is drawn into the cleaned pipet by means of a pipet bulb; a variety of types are available. **Never use mouth suction to fill a pipet.** The liquid is drawn above the graduation mark, and the index finger is placed firmly on the opening at the top of the pipet. By carefully moving the index finger, the liquid is then allowed to drain slowly until the meniscus is exactly on the desired graduation mark. In the case of the volumetric pipet, the tip should be kept in contact with the side of the receiving container as the liquid drains into the container. A final drop is removed by touching the side of the container with the tip of the volumetric pipet. By design, a small amount of liquid will remain in the "TD" volumetric pipet; it is now "blown out." See Figure 0-9. With a "TC" serological pipet, when the entire contents is to be delivered, the last drop is "blown out" of the tip.

Besides the serological type, there is a second type of graduated pipet, the Mohr pipet. It is not graduated to the tip but rather to a line above. It is never drained completely, just from one graduation to another.

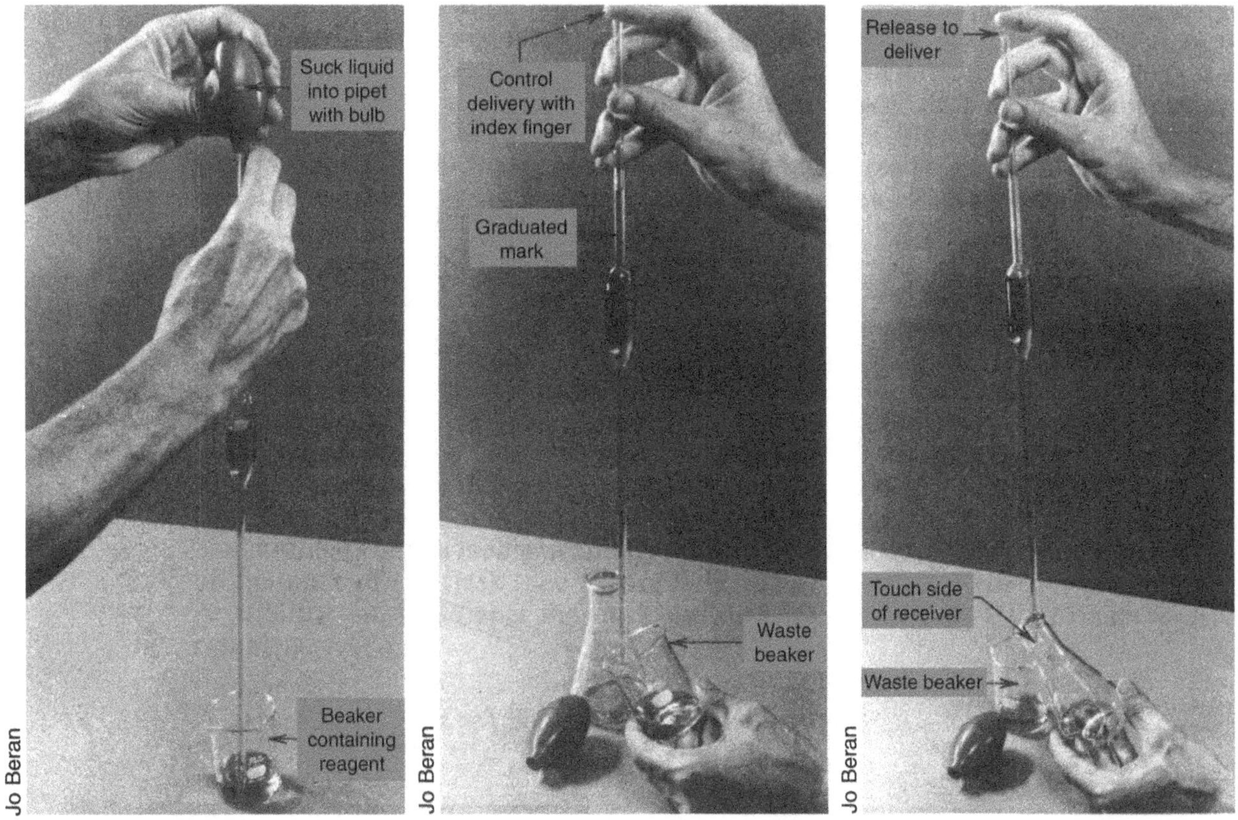
Suck liquid into pipet with bulb
Beaker containing reagent
Jo Beran
Control delivery with index finger
Graduated mark
Waste beaker
Jo Beran
Release to deliver
Touch side of receiver
Waste beaker
Jo Beran

Figure 0-9 Delivering a liquid from a volumetric pipet.

c. Burets

1. Description A buret is a hollow, narrow, cylindrical tube labeled with graduation marks and equipped with a regulating device (usually a stopcock) for controlling the flow of liquid from the tip. Since it has a narrow, uniform diameter, it is possible to deliver small quantities of liquid with a high degree of accuracy. The graduation marks on a buret begin with zero near the tip and increase down the buret. The final graduation mark is near, but not at, the bottom of the buret.

2. Preparation of the Buret Clean the buret with detergent and water before use. Do not pour solid detergent into the buret since it is apt to block the narrow tip; instead, dissolve a small amount of detergent in water and pour the solution into the buret. Use a long-handled buret brush to scrub the inside of the buret, but use extreme care to prevent the wire handle from scratching the inside walls of the buret. Rinse thoroughly with tap water, allowing the water to drain through the tip.
Close the stopcock and fill with deionized water. Place a beaker or flask beneath the tip and allow the water to drain. Check to see that the water flows freely and that there are no leaks at the stopcock. Water droplets should not cling to the walls of the buret. Close the stopcock and add about 5 mL of the reagent solution (**titrant**). Tilt the buret and roll it so the titrant comes in contact with the entire inner surface. Place the buret in an upright position and allow the titrant to drain into a beaker or flask. Discard this titrant since it has been diluted by residual rinse water left in the buret.

Ask your instructor for assistance with a leaking stopcock.

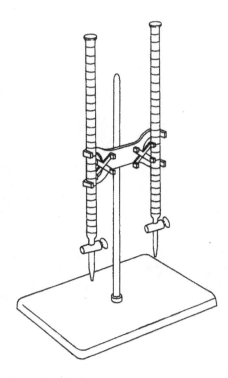

Figure 0-10 Proper method for supporting burets.

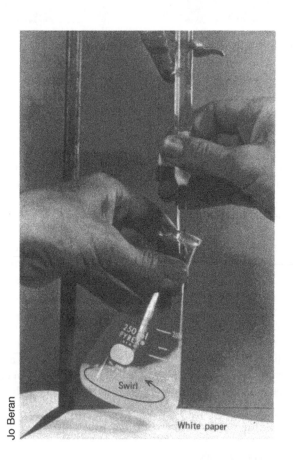

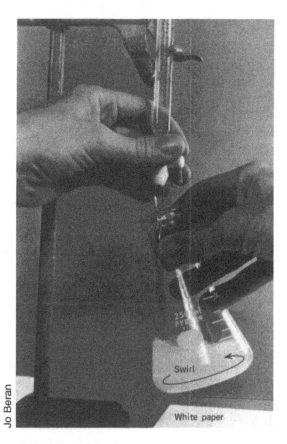

Figure 0-11 Correct titration procedure.

Support the buret with a buret clamp. See Figure 0-10. Close the stopcock and fill to the top with the titrant to be used. Allow the titrant to drain through the tip until there is a solid column of liquid extending down to the tip, completely free of air bubbles.

3. Using the Buret The process of determining the amount of solution required to react with a given amount of sample is called a **titration**. The sample is placed in a receiving vessel (usually an Erlenmeyer flask). An **indicator**, a substance that changes color when the endpoint is reached, is added to the receiving flask. Place the flask on a sheet of white paper so that you can clearly see the color change of the endpoint.

Record the initial buret reading to the nearest 0.01 mL. It is not necessary for the initial reading to be exactly 0.00 mL—an initial reading of 2.45 mL or 3.10 mL is every bit as accurate.

Manipulate the stopcock with your left hand and swirl the flask with your right hand. (A left-handed person may find it more comfortable to switch hands.) See Figure 0-11.

Dispense the titrant rapidly at first. You will notice that a color will appear momentarily at the spot in which the titrant comes in contact with the sample in the flask. As you near the endpoint, the indicator's color will persist over a longer interval of time. At this point, slow the addition of the titrant so that you do not add too much and go past the endpoint. While swirling the flask, add the titrant slowly, drop by drop until one drop finally produces a color which will not fade for at least 30 seconds. Allow 30 seconds for the titrant to drain from the walls of the buret; then read the buret to the nearest 0.01 mL. The volume of titrant used is the difference between initial and final readings.

d. Volumetric Flasks

These long-necked flasks are used to prepare solutions of specific concentrations. In practice, a measured amount of solute is placed in the clean flask. Add solvent until the flask is about half-full; it is much easier to swirl the solution and dissolve the solute when the flask is only half-full. Also, the volume of the solution often increases or decreases as the solute dissolves and, therefore, the solute must be totally dissolved before final filling of the flask to the graduation mark. Swirl the contents of the half-filled flask to completely dissolve the solute. Additional solvent is then added to the flask, the last few milliliters with a medicine dropper, until the meniscus reaches the graduation mark on the neck of the flask. See Figure 0-12. Be especially careful not to add solvent beyond the mark. Stopper the flask (most have a glass stopper) and turn it upside down 10-12 times to insure complete mixing.

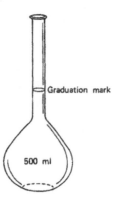

Figure 0-12 Volumetric flask.

4. The Laboratory Balance

When the mass of an object must be measured with an uncertainty of 0.01 g, a top-loading balance is frequently used (see Fig. 0-13). For higher precision (±0.001 g or ±0.0001 g) other electronic balances are available and are called "analytical" balances (see Fig. 0-14).

Some balances have a "tare" feature that is useful when finding the mass of a substance being added to a container. Enabling the tare function causes the balance display to read zero, even when a container is on the pan of the balance. When a substance is added to the container, the balance displays only the mass of the substance, not the combined mass of the substance and container.

You instructor will supply specific directions for the use of the balance in your laboratory. The following rules apply to every balance:

1. Never place a reagent directly on the weighing pan. Use either weighing paper or an appropriate container, such as a beaker.
2. Clean up all spilled materials.
3. Never return excess reagent to its original bottle. Your instructor will advise you as to its disposal.
4. If the balance appears to be out of adjustment, inform your instructor. Never attempt to make adjustments on your own.
5. Do not move the balance from one location to another. The balance must be readjusted after movement and movement may permanently damage the balance.

Figure 0-13 Top-loading balance.

Figure 0-14 High-precision balance.

5. Laboratory Heat Sources

a. Bunsen Burner

1. Description

In 1855 the German chemist Robert Bunsen popularized a burner that mixed air with flammable gas to produce a hot, smokeless flame. Today, all laboratory burners are commonly referred to as Bunsen burners even if they are modifications of his design. The original Bunsen burner design did not have a gas adjustment valve. Most modifications, such as the Tirrill burner, have a gas adjustment screw at the base for controlling the supply of gas. The gas supply of the Bunsen burner must be controlled at the gas outlet valve on the laboratory bench—a less precise adjustment method. The supply of air entering the barrel of the burner is regulated either by adjusting a sleeve at the base or by screwing the barrel up or down on the base, depending upon which burner is used. See Figures 0-15 and 0-16.

The principal component of natural gas used in most laboratories is methane (CH_4). Because methane is odorless, a gaseous substance possessing a distinct odor is added to the methane so that gas leaks may easily be detected. When methane is burned in a sufficient supply of air, it combines with oxygen (O_2) to form carbon dioxide gas (CO_2) and water vapor (H_2O). The hot flame has two distinct blue cones and is nonluminous (does not produce light). See Figure 0-17.

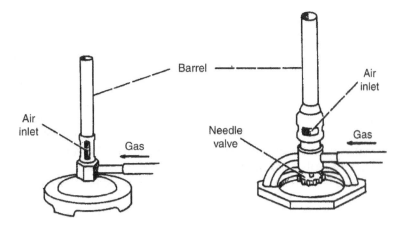

Figure 0-15 Bunsen burner. **Figure 0-16** Tirrill burner.

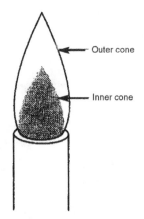

Figure 0-17 The flame of a properly adjusted Bunsen burner has two blue cones.

With an insufficient supply of air, the combustion of methane is incomplete—carbon (C) and carbon monoxide (CO) are produced. The tiny, hot, glowing pieces of solid carbon become incandescent and produce a yellow, luminous (light-producing) flame. Objects heated with a luminous flame become coated with black soot (unburned carbon). The carbon monoxide produced by incomplete burning of methane is a toxic, colorless, odorless gas.

2. Operation

Attach rubber tubing from the burner to the gas outlet on the lab bench. Fully close the air inlet on the burner, then open it slightly. Fully close the gas control valve on the burner, then open it slightly. Fully open the gas outlet on the lab bench. Bring a lighted match or friction lighter to the side of the burner at the top of the barrel and ignite the gas. Adjust the burner's gas control valve and air inlet for your experiment. For most laboratory operations, the flame should have two distinct blue cones and be about 8 cm (3 inches) tall. The flame should burn with a very slight "roar." Excessive "roaring" indicates the air inlet is open too far—this causes the gas to burn inside the barrel of the burner and the barrel will get hot.

b. Hotplates (Hotplate/Magnetic Stirrer)

1. Description

Electrical heat sources are another very common means used for heating in the laboratory. The most common type is probably a hotplate that may be equipped with a magnetic motor to provide even stirring during heating. One common example is shown in Figure 0-18. Those equipped with the stirring motor will have two controls. One dial controls the heating rate and the other dial the stirring rate. Most have a chemically resistant heating surface such as porcelain. Frequently the porcelain surface is white when cool and a yellowish color when hot so that visual observation will indicate the relative temperature of the surface. Because there is no flame associated with hotplates, they are somewhat easier and safer to use. These are often capable of providing temperatures up to 500°C if the heating is done without a water bath. Whether or not the container to be heated is placed directly on the heating surface, in a water bath, surrounded by an aluminum block or another method depends on what is being heated as well as the temperature needed.

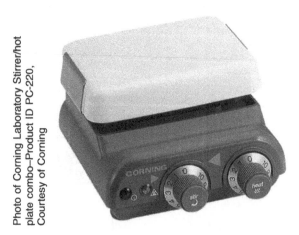

Photo of Corning Laboratory Stirrer/hot plate combo–Product ID PC-220, Courtesy of Corning

Figure 0-18 A laboratory hotplate/magnetic stirrer.

2. Operation

A flat wire gauze may be placed on the cool surface of the hotplate and the container or the bath that is to be heated than placed on the center of the gauze but this is optional. A magnetic stir bar or boiling stone should be placed in any liquid to be heated. Never add a boiling stone or stir bar to a hot liquid as this may cause bumping and dangerous spattering of the liquid out of the container. If the magnetic stir bar is used, turn the stirring motor on to provide even spinning of the stir bar. Boiling stones and the stir bar prevent hot spots from developing as the liquid is heated which can cause bumping of the liquid. Bumping can cause the liquid to be ejected from the container or cause the container to crack or tip over. When ready to start heating, the heat control is turned up slowly to obtain the desired temperature without overheating.

6. Heating Liquids

Whether a Bunsen burner or hotplate is used depends on the temperature needed. Bunsen burners can provide temperatures higher than those attainable with a hotplate but are not recommended when using flammable chemicals.

a. In a Test Tube

1. Using a Bunsen Burner

It is advised that burners never be used to heat flammable liquids or volatile solids. The test tube should not be more than one-third full. Clamp the tube at the top so as to provide a large heating surface. Incline the test tube slightly, and direct the flame at the top of the liquid, not at the bottom. Move the test tube constantly. See Figure 0-19.

CAUTION: Do not point the open end of a test tube toward another person—the liquid may suddenly shoot out if the test tube is not properly heated. Do not heat a closed or stoppered container—pressure will build up in the container. Never heat graduated cylinders, bottles, watch glasses, or other thick-walled glassware—they will break.

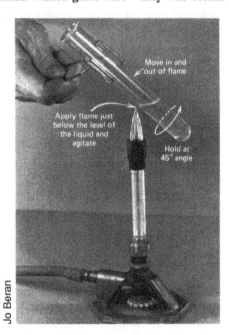

Figure 0-19 Heating a liquid in a test tube.

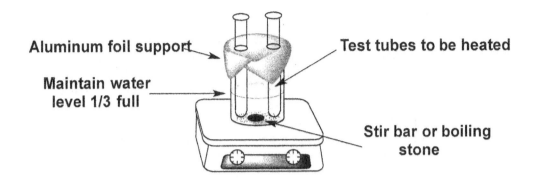

Figure 0-20 Water bath containing test tubes.

2. Using a Hotplate

Test tubes are usually heated in a bath (commonly water). The test tube is supported in the bath to prevent tipping. A simple apparatus for doing this is shown in Figure 0-20. The beaker is partially filled with water, a boiling stone or stir bar is added, and a piece of aluminum foil is placed over the top of the beaker and a small hole is made in the foil with the end of a spatula. The test tube is fitted through the hole and is supported by the foil. The stirring motor, if applicable, is turned on to obtain even stirring and then the heat control is adjusted to obtain the desired temperature.

b. In a Beaker or Flask

1. Using a Bunsen Burner

Set up a ring-stand with iron ring and wire gauze as illustrated in Figure 0-21. A second iron ring prevents the beaker from falling off the wire gauze. Place a boiling stone in the liquid [black (SiC) type if liquid is acidic] to avoid violent bumping or spattering of the liquid. Alternatively, use a stirring rod and then position the flame beneath the tip of the stirring rod.

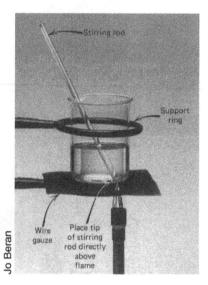

Figure 0-21 Proper method for heating a liquid in a beaker.

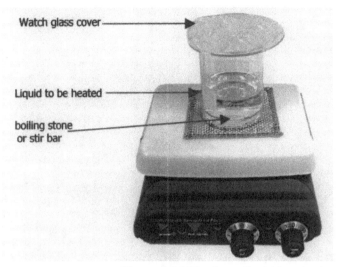

Watch glass cover

Liquid to be heated

boiling stone
or stir bar

Jo Beran

Figure 0-22 Heating a liquid in a beaker.

2. Using a Hotplate

Add a magnetic stir bar or boiling stone to the liquid [black (SiC) type if liquid is acidic] to be heated in a beaker or flask. Usually a watch glass cover is placed on the container to prevent evaporative loss of the liquid while not giving an airtight seal. Place the container on the center of the wire gauze (if used) on the hotplate. Start the stirrer motor, if present. Again slowly turn the heating control to obtain the desired temperature. This set up is shown in Figure 0-22.

7. Evaporating Liquids

CAUTION: Do not use a flame near flammable liquids. Use a fume hood if noxious odors are likely to be produced.

1. Using a Bunsen Burner

a. Over an Open Flame

Nonflammable liquids may be evaporated from an evaporating dish using a gentle, "cool" flame as illustrated in Figure 0-23. When the quantity of liquid remaining in the evaporating dish becomes very small, remove the burner to avoid spattering.

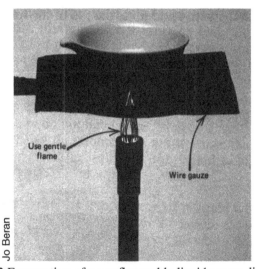

Use gentle flame

Wire gauze

Jo Beran

Figure 0-23 Evaporation of a nonflammable liquid over a direct flame.

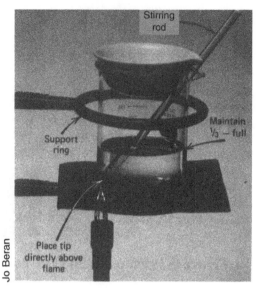

Figure 0-24 Evaporation of a nonflammable liquid over a steam bath.

b. Over a Steam Bath

Nonflammable liquids may also be evaporated from a steam bath. The evaporating dish is placed over a beaker of boiling water. See Figure 0-24. The steam bath provides a moderate, uniform distribution of heat that prevents burning, scorching, or overheating.

c. Using a Hotplate

Usually liquids are evaporated using a water bath on a hotplate. The beaker that will serve as the bath is partially filled with water and a boiling stone or magnetic stir bar is added. The container, usually an evaporating dish, with the liquid to be evaporated is placed on top of the beaker. The spout of the beaker should allow air to enter or water vapor to leave the beaker so that an airtight system is not set up. See Figure 0-25. The stirring motor, if present, is turned on for even stirring and the heat control turned up to allow the water to boil. The water vapor carries the heat to the bottom of the container causing the liquid to evaporate. Do not allow the water bath to go to dryness nor the container being heated. When the liquid has nearly evaporated from the container, with tongs, remove the container with the sample and place it on another wire gauze to finish evaporating and to subsequently cool on the bench top.

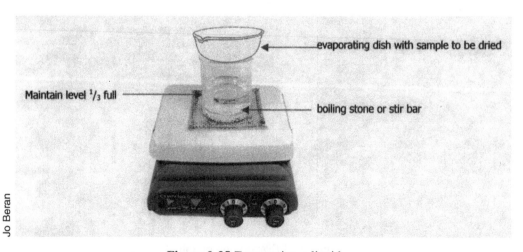

Figure 0-25 Evaporating a liquid.

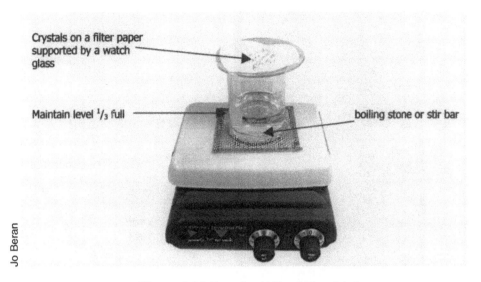

Crystals on a filter paper
supported by a watch
glass

Maintain level $^1/_3$ full

boiling stone or stir bar

Jo Beran

Figure 0-26 Crystals on filter being dried.

Solids isolated by filtration can be dried in a similar manner by opening the filter paper and placing it on a watch glass. The watch glass can then be placed on top of the water bath. See Figure 0-26.

8. Filtration

Gravity: Figure 0-27 illustrates how to fold a filter paper and arrange the paper for filtration. The paper is first folded in half. Then fold the paper again, not quite in perfect quarters. The two folded edges should not quite touch; the second edge should be about 3 mm (one-eighth inch) from the first edge. Do not crease the folds too sharply or a hole may develop at the point of the paper. A small corner is torn from the smaller section. The tear enables the paper to make a tight seal against the funnel. The paper is then opened to form a cone and inserted into the funnel. Moisten the paper with a small amount of solvent (usually water) and press the top of the filter paper against the funnel to make a tight seal.

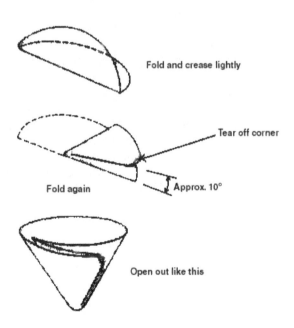

Fold and crease lightly

Tear off corner

Fold again

Approx. 10°

Open out like this

Figure 0-27 Preparing filter paper for filtration.

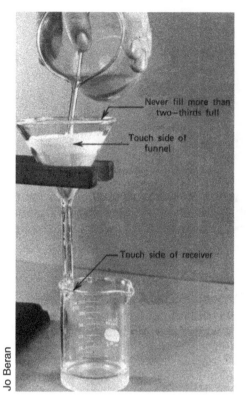

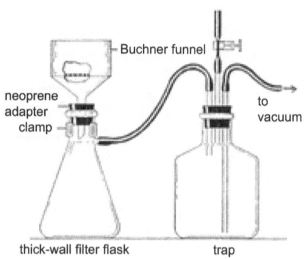

Figure 0-28 Proper filtration set-up.   **Figure 0-28A**.

The funnel should be supported with a funnel rack or with a wire triangle on an iron ring. To avoid splashing and to increase filtration speed, the tip of the funnel stem should touch the side of the receiving vessel. See Figure 0-28. Never fill the funnel more than two-thirds full of solution.

Vacuum: To speed up slow filtering mixtures, vacuum filtration is often useful. A special type of funnel, usually made of porcelain, is used. It is placed in a Neoprene rubber gasket adapter and then placed in a filter flask. These are made with thick glass walls and have a side arm for attachment to a heavy walled bottle or filter flask that acts as a trap to prevent filtrates that foam or are of excessively large volume from entering the vacuum system. The trap in turn is connected to a vacuum source, commonly an aspirator attached to a water faucet or a house vacuum system.

There are two types of funnels used, the conical Hirsch funnel or the Buchner funnel, which has cylindrical upper walls. For use, a disc of filter paper is placed on the perforated disc incorporated in the funnel. It is important that it be sized to lay flat, just covering the funnel perforations. If filtering an aqueous suspension, the filter paper disc is moistened with a few drops of water to obtain a good seal. It is very important that both the filter flask and the trap are securely fastened to ring stands; if they were to tip over when evacuated they could implode and throw glass and chemicals around. An example of vacuum filtration equipment is shown in Figure 0-28A.

9. The Laboratory Thermometer

Laboratory thermometers can be of the liquid-filled type. For routine use, most laboratories use spirit-filled thermometers, filled with hexane, mineral oil, or alcohol, which is dyed red for visibility.

Laboratory liquid-filled thermometers do not have to be "shaken down" before use. In fact, shaking or dropping a liquid-filled thermometer is apt to cause the liquid column to separate, meaning there is a gap in the liquid. If the liquid column is separated, give the thermometer to your instructor for repair.

Digital thermometers are replacing liquid-filled thermometers. In general, they are more precise and many models are quite inexpensive. The dials of these thermometers usually display the temperature of the sample to ±0.1 or 0.2°C. They have a fairly rapid equilibration time and are not subject to the column separation problems and breakage of liquid-filled thermometers. They also cover a wider temperature range than that of many liquid-filled thermometers. They do require a battery. Operation is straightforward and usually only requires pushing a button on the face of the thermometer. An example of a digital thermometer is shown in Figure 0-29.

Courtesy of DeltaTRAK Inc.

Figure 0-29 A simple inexpensive digital thermometer.

You might need to measure the temperature of a liquid while you are heating it. The thermometer should be held in the center of the liquid. Do not rest the thermometer on the bottom of the beaker since the glass beaker may be a different temperature than the liquid it contains.

A thermometer must be allowed time to adjust to the temperature of the substance being measured. A laboratory thermometer usually responds rapidly, perhaps within 30 seconds depending on how much the substance's temperature differs from the thermometer's initial temperature. Watch the liquid column move; after it stops moving wait at least another 30 seconds before taking a temperature reading. For digital thermometers, monitor the display until it reaches a steady value.

Fundamental Laboratory Operations

Experiment **1**
Measurement

Reference: *General, Organic, and Biological Chemistry: An Integrated Approach, 4th ed., Chapter 1.*

Purpose: To familiarize you with the measurement process and use of the balance, graduated cylinder, thermometer, and urinometer.

Introduction

MEASUREMENT SYSTEMS

Nearly all the world's people use a set of measurement units called *Le Système International de d'Unités*, commonly referred to as the SI system. In 1960, the SI system was developed to replace the older metric system. The SI system is popular because it is a decimal number system. Every unit is ten times the size of the next smaller unit; for example there are 10 millimeters in 1 centimeter. The familiar English system of measurements is not a decimal system—there are 4 quarts in 1 gallon, 12 inches in 1 foot, and 16 ounces in 1 pound.

The SI system uses a small number of base units (such as the meter) and a set of prefixes (such as kilo-, centi-, and milli-) which can be used with any base unit. Chemists continue to use a few older metric units such as the liter and the milliliter that are not part of the SI system. A volume of 1 liter (1 L) equals 1 cubic decimeter (1 dm^3) and 1 milliliter (1 mL) is equal to 1 cubic centimeter (1 cm^3).

Measured quantities almost always consist of two parts: a number and a unit. You must get in the habit of writing units with your measured numbers. Don't write a length as 5; write it as 5 mm, 5 cm, or 5 m.

MEASUREMENTS ARE ALWAYS ESTIMATES

Much of the work in a chemistry laboratory involves measurements. Unlike counting, which can be exact, measurements are never exact but are always estimated quantities. All measurements are comparisons to some standard. No matter how precise the measurement instrument, the last digit measured must be an estimate. Obviously, some instruments make better estimates than others and experienced investigators make better measurements than inexperienced investigators.

In making length measurements with a 12-inch ruler, the familiar English system divides each into halves, quarters, eighths, and sixteenths of an inch. Because it is not a decimal number system, measurements with the English ruler are usually done by finding the mark closest to the length being measured. In Figure 1-1, the length is estimated as $1\frac{7}{16}$ inch.

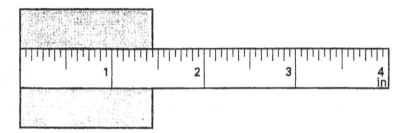

Figure 1-1 Measure to the nearest mark ($1\frac{7}{16}$ inch) when using an English ruler.

In contrast, because the SI system is a decimal system, each centimeter is divided into tenths of a centimeter rather than sixteenths of a centimeter. A length of 0.1 cm is the same as 1 mm.

Most measurements are read from a scale such as the one in Figure 1-2. A scale such as this is commonly used on instruments that measure length (a ruler), mass (a balance), temperature (a thermometer), and volume (volumetric glassware such as a graduated cylinder, a buret, or a syringe). The scale has graduation marks and some of the graduation marks have printed numbers. Usually there are 10 graduation marks between the printed numbers but some scales have only 5 graduation marks between the printed numbers.

In using the scale, you should imagine additional graduation marks that are between the marks printed on the scale. As a rule of thumb, you should try to imagine 10 additional marks between the printed marks. The length in Figure 1-2 is between two graduation marks, 4 cm and 5 cm. The 10 imaginary marks are 4.0, 4.1, 4.2, 4.3, 4.4, 4.5, 4.6, 4.7, 4.8, and 4.9 cm. The length can be estimated as 4.2 cm.

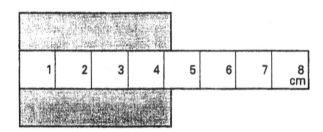

Figure 1-2 The reading (4.2 cm) is between two graduation marks.

When the reading appears to lie exactly on a printed graduation mark, add a zero to the reading to indicate the imaginary mark was zero. The length in Figure 1-3 should be estimated as 4.0 cm.

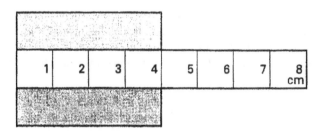

Figure 1-3 The reading (4.0 cm) is on a printed graduation mark.

Each of these last two measurements used the same measurement instrument, one where the smallest graduation mark was 1 cm. Each measurement was estimated to the nearest 0.1 cm. Thus, each measurement has 1 digit after the decimal point. The measurement in Figure 1-3 is 4.0 cm rather than 4 cm. The measurement of 4.0 cm indicates the measurement is estimated ±0.1 cm whereas a measurement of 4 cm indicates a less precise estimate of ±1 cm.

When the printed marks are very close together it may not be possible to imagine 10 additional marks and you may be able to imagine only 1 additional mark. An investigator can decide whether the reading appears to lie ON a graduation mark or BETWEEN two closely spaced graduation marks. The graduation marks in Figure 1-4 are every 0.1 cm (1 mm). Although it is possible to imagine 10 additional marks, inexperienced investigators might imagine only 1 additional mark. The length in Figure 1-4 is BETWEEN 4.0 cm and 4.1 cm. If it were ON the 4.0 cm mark we could write the measurement as 4.00 cm. If it were ON the 4.1 cm mark we could write the measurement as 4.10 cm. Since the length appears to be BETWEEN 4.0 and 4.1 cm, we could use an imaginary mark to get the measurement 4.05 cm. This measurement has been estimated to the nearest 0.05 cm. With experience it is possible to make better estimates for the imaginary marks.

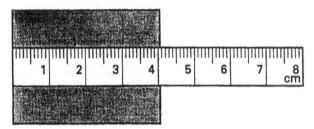

Figure 1-4 When the printed marks are very close together, you can use 1 intermediate imaginary mark. The reading is 4.05 cm.

Generally, no matter how closely spaced the graduation marks, you should estimate the measurement by imagining additional marks between the printed marks on the scale. This measurement estimation technique should be used for all measurement scales whether found on a ruler, balance, thermometer, or graduated cylinder. Even many fundamental laboratory measuring devices are becoming digital. The numerical display of such devices directly gives the measurement to the inherent uncertainty. In such cases, the observer does not have to make the estimate. For example, if the readout of a balance is 10.00 g, the inherent estimation (uncertainty) is ±0.01 g.

USE OF THE THERMOMETER

Review Section 9 of Fundamental Laboratory Operations for procedures in using a laboratory thermometer.

USE OF THE BALANCE

Review Section 4 of Fundamental Laboratory Operations, page 16, for general procedures in using a laboratory balance. Your instructor will supply specific directions for use of the balance in your laboratory. Remember that this measuring instrument is a balance, not a scale.

VOLUMETRIC GLASSWARE

Be thoroughly familiar with Section 3 of Fundamental Laboratory Operations, pages 10-15, for descriptions and proper uses of graduated cylinders and volumetric flasks. All glassware must be thoroughly cleaned before use and cleaned again before storage.

DENSITY

Density is the mass of a substance per unit volume of that substance. In equation form:

$$\text{density} = \frac{\text{mass}}{\text{volume}} \quad \text{or} \quad d = \frac{m}{v}$$

The density of a solid or liquid is expressed in units of grams per cubic centimeter (g/cm^3) or the equivalent grams per milliliter (g/mL). Density is a constant at a particular temperature and does not depend on the size of the sample. Water, for example, has a density of 1.00 g/mL at 4°C. At any other temperature liquid water is slightly less dense, but still very close to 1.00 g/mL.

When determining the density of a liquid, mass and volume can be obtained by direct measurements; a balance is used for measuring mass, and a graduated cylinder is used for measuring volume.

If a solid is insoluble in water, its volume can be indirectly determined by measuring its water displacement. Water is poured into a graduated cylinder and its volume carefully measured. The graduated cylinder is tilted and the solid is allowed to submerge beneath the surface of the water. The change in volume is the volume of the solid. See Figure 1-5.

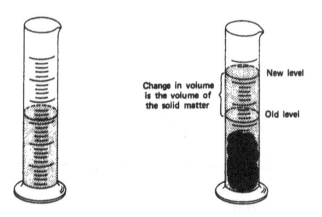

Figure 1-5 Measuring the volume of a solid by water displacement.

If a solid object floats on water, its density must be less than water's density. If the solid's density is greater than water's density, the solid will sink in water.

SPECIFIC GRAVITY

The specific gravity of a liquid or solid is the density of that substance divided by the density of water (measured at the same temperature).

$$\text{specific gravity} = \frac{\text{density of the substance}}{\text{density of water}}$$

Recall that water's density is 1.00 g/mL at 4°C. If the density of the substance is also expressed in units of g/mL, the specific gravity of the substance will be the same number as the density but without any units—the units cancel. For example, ethyl alcohol has a density of 0.789 g/mL and a specific gravity of 0.789. Specific gravity is one of very few quantities without any units.

The specific gravity of urine is valuable in medical diagnosis. The specific gravity indicates the concentration of solids such as urea, phosphates, chlorides, proteins, and sugars which are dissolved in the urine. The normal specific gravity of urine ranges from 1.005 and 1.030, with most samples falling between 1.010 and 1.025.

If a person's urine produces a continuously low specific gravity reading, this may indicate diabetes insipidus, a disease caused by impairment of an antidiuretic hormone. Low specific gravity readings may also indicate that a kidney has lost its ability to concentrate urine because of tubular damage.

High urine specific gravity readings occur in patients with diabetes mellitus, adrenal inefficiency, hepatic diseases, and congestive cardiac failure. The specific gravity of urine may be elevated due to excessive loss of water caused by sweating, fever, vomiting, or diarrhea.

The specific gravity of a liquid can be measured with a hydrometer. The **hydrometer** consists of two pieces—a jar to hold the liquid, and a float. The float is weighted to make it partially sink in the liquid. The amount the float sinks is determined by the liquid; the float sinks less in liquids with a high specific gravity. The float has a thin stem with a calibrated scale. The specific gravity reading is made at the position the surface of the liquid meets the scale. A **urinometer** is a hydrometer especially designed and calibrated for urine. See Figure 1-6.

Figure 1-6 A urinometer. The reading is made where the float meets the surface of the liquid.

Typical urinometer scales are calibrated from 1.000 to 1.060. Because of a lack of space on the thin stem of the float, decimal points are often left out; a reading of 1028 is really 1.028. The graduation marks on a urinometer are very close together; so close that it is impractical to use imaginary marks between the printed marks. See Figure 1-7.

Figure 1-7 The urinometer reading is 1.028. Notice that the numbers increase going **down** the stem of the float.

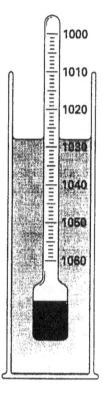

For accurate work, the temperature of the liquid should be taken. A small correction to the specific gravity can be applied for every degree the liquid is above or below 4°C. This experiment will not make a temperature correction.

Procedure

Note: Enter all data and answers on the **Report Sheet**.

I. LENGTH

Use an English-metric ruler to make the following measurements as accurately as you can. *Make sure you write the units of each measurement along with the number.* Use a pen or pencil to check off each step as you complete the step.

1. Measure the length of the smallest test tube in your locker in millimeters and in inches.

2. Measure the dimensions of the front cover of this lab manual in English units and in SI units. Calculate its area using each set of units.

II. TEMPERATURE

1. Fill a beaker with cold tap water. Measure the temperature of the cold tap water.

2. Fill a beaker with hot tap water. Measure the temperature of the hot tap water.

III. MASS

1. Measure the mass of a piece of weighing paper.

    [If your balance has a "tare" function, you may use the following procedure: Place the paper on the pan of the balance and, if you choose, press "tare." On the Report Sheet for the "mass of the weighing paper" you may write "TARE" or "0.00 g" rather than the mass.]

2. Place 5 aspirin tablets on the weighing paper and record the mass of the paper and tablets.

    [If you have chosen to use the "tare" function of your balance, write "TARE" or "0.00 g" for the "mass of the weighing paper and 5 aspirin tablets." Record the "mass of 5 aspirin tablets" and proceed to Step 5.]

3. If the "TARE" function is not available, calculate the mass of the 5 aspirin tablets.

4. Calculate the average mass of 1 aspirin tablet.

5. Follow your instructor's directions for the disposal of the aspirin tablets.

IV. VOLUME

1. Examine the graduation marks of a 50-mL graduated cylinder. Look carefully, some graduation marks may be missing. On the Report Sheet draw each graduation mark below the 10-mL mark. Label all the graduation marks with their volume.

2. Examine the graduation marks of a 10-mL graduated cylinder. On the Report Sheet draw each graduation mark below the 3-mL mark. Label all the graduation marks with their volume.

3. Fill a standard (16 × 125 mm) test tube to the brim with tap water and pour the contents into a 50-mL graduated cylinder. Measure the volume to the nearest 0.1 mL. Remember to read the graduated cylinder at the bottom of the meniscus.

4. Fill a small (10 × 75 mm) test tube to the brim with tap water and pour the contents into a 10-mL graduated cylinder. Measure the volume to the nearest 0.1 mL.

5. Half fill a small beaker with tap water. Use a medicine dropper to transfer 1 mL of water to a 10-mL graduated cylinder. Count the number of drops in 1 mL.

V. DENSITY

A. Density of Water

1. Measure the mass of a dry 50-mL graduated cylinder.

2. Pour exactly 50.0 mL of deionized water into the graduated cylinder. Use a medicine dropper to adjust the meniscus to the 50-mL mark.

3. Measure the mass of the graduated cylinder with the water.

4. Calculate the mass of the water.

5. Calculate the density of the water. Remember to include units with each quantity.

B. Density of a Sugar-Water Solution

1. Measure the mass of a clean, dry 100-mL volumetric flask with its cap (or stopper).

2. Using a piece of weighing paper as in Part III, crease it lightly, tare it, and weigh unto it 10.00 g of sucrose (table sugar). Then carefully slide the sucrose into the flask from the creased paper sheet, being careful to avoid spilling.

3. Fill the flask halfway with deionized water. Swirl the flask until the sugar dissolves.

4. Fill the flask to the 100-mL mark with deionized water. Use a medicine dropper to carefully add the last milliliter of water. Remember that the bottom of the meniscus should be at the 100-mL mark.

5. Tightly cap the flask and invert it about 15–20 times to obtain a uniform solution. Measure the mass of the capped flask and its contents.

6. Calculate the mass of the sugar-water solution.

7. Calculate the density of the sugar-water solution.

8. **Save the solution for use in Part VI.**

C. Density of a Rubber Stopper

1. Measure the mass of a No. 1 solid rubber stopper.

2. Pour 20–25 mL of tap water into a 50-mL graduated cylinder. Record the volume to the nearest 0.1 mL.

3. Tip the graduated cylinder and slowly slide the stopper into the graduated cylinder until it is entirely beneath the surface of the water. Place the graduated cylinder in an upright position and record the volume to the nearest 0.1 mL.

4. Calculate the volume of the rubber stopper and the density of the rubber stopper.

D. Density of an Unknown Solid

Obtain an unknown solid from your instructor and determine its density.

VI. SPECIFIC GRAVITY

1. Use a urinometer to measure the specific gravity of deionized water. Fill the urinometer jar with deionized water to 2 cm from the top.

2. Gently add the float to the water with a slight spinning motion. Avoid wetting the float stem above the liquid line. Make sure the float is actually floating and not resting on the bottom of the jar.

3. Read the scale of the urinometer float at the lowest portion of the meniscus of the water. Keep the float away from the sides of the jar while reading.

4. Use the urinometer to measure the specific gravity of the sugar-water solution prepared in Part V. B.

5. Discard the sugar-water solution in the sink. Rinse the volumetric flask and the urinometer several times with tap water and once or twice with a few milliliters of deionized water.

6. Wash your hands before leaving the laboratory.

Name _____ Section _____ Date _____

Prelab Questions for Experiment **1**

1. Record the following measurements.

a.
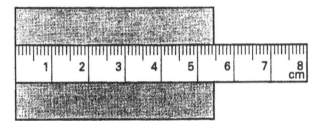

b.
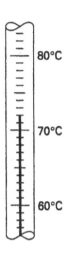

c.
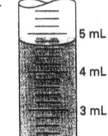

d.
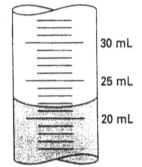

2. A 15-gram solid object is dropped into a graduated cylinder containing 35 mL of water. The water level rises to the 45-mL mark.

   a. What is the density of the solid? Show your calculations.

   b. What is the specific gravity of the solid?

3. a. What is a meniscus?

   _____

   b. Explain the proper technique for reading a meniscus.

   _____

   _____

4. When preparing a solution with a volumetric flask, why is the solute first dissolved in a small quantity of solvent before filling the flask to the mark?

   _____

   _____

Report for Experiment 1

I. LENGTH

Remember to include units with each quantity.

1. a. Length of a small test tube _____ mm _____ in

   b. How many millimeters are there in an inch?

   _____

2. a. Dimensions of front cover in English units _____ × _____

   b. Dimensions of front cover in SI units _____ × _____

   c. Area of front cover in English units _____

   d. Area of front cover in SI units _____

II. TEMPERATURE

1. Temperature of cold tap water _____

2. Temperature of hot tap water _____

III. MASS

1. Mass of weighing paper and 5 aspirin tablets _____

2. Mass of weighing paper _____

3. Mass of 5 aspirin tablets _____

4. a. Average mass of 1 aspirin tablet

   b. If the mass of aspirin in each tablet is 5 grains (1 grain = 0.065 g), calculate the mass in grams of aspirin in 1 tablet.

Calculations: _____

   c. Calculate the percentage of aspirin in each tablet.

$$\% \text{ aspirin} = \frac{\text{mass of aspirin}}{\text{mass of tablet}} \times 100$$

IV. VOLUME

1. Draw and label the graduation marks on a 50-mL graduated cylinder below the 10-mL mark.

2. Draw and label the graduation marks on a 10-mL graduated cylinder below the 3-mL mark.

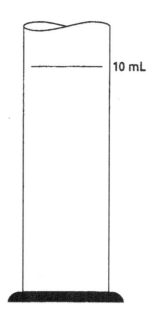

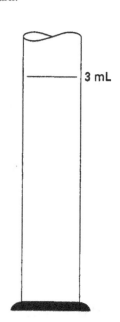

3. Volume of standard test tube   _____

4. Volume of small test tube   _____

5. Number of drops in 1 mL   _____

V. DENSITY

A. Density of Water

1. Mass of empty graduated cylinder   _____

2. Mass of graduated cylinder and water   _____

3. Mass of water   _____

4. Volume of water   _____

5. Density of water   _____

Calculations:

6. Why might the experimentally determined density of water differ from its expected density?

B. Density of Sugar-Water Solution

1. Mass of volumetric flask and solution   _____

2. Mass of empty volumetric flask   _____

3. Mass of solution   _____

4. Density of solution   _____

     Calculations:

5. How does your measurement of the density of the 10% w/v sucrose (table sugar) solution compare with the density value reported in the *Handbook of Chemistry and Physics*? [See the index in the back of any available edition and look for: Density, organic substances, selected, varying concentration in aqueous solution, table, D-220-267 (which may vary from one edition to another).]

_____

_____

C. Density of a Rubber Stopper

1. Mass of rubber stopper   _____

2. Volume of water and rubber stopper   _____

3. Volume of water   _____

4. Volume of rubber stopper   _____

5. Density of rubber stopper   _____

     Calculations:

D. Density of an Unknown Solid

    Unknown Number   _____

1. Mass of solid   _____

2. Volume of water and solid   _____

3. Volume of water   _____

4. Volume of solid   _____

5. Density of solid   _____

    Calculations:

VI. SPECIFIC GRAVITY

1. a.   Specific gravity of water   _____

   b.   How does the specific gravity of water measured with the urinometer compare with the density calculated in Part V. A?

         _____

         _____

2. a. Specific gravity of sugar-water solution   _____

   b.   How does the specific gravity of the sugar-water solution measured with the urinometer compare with the density calculated in Part V. B?

         _____

         _____

   c.   If your sugar-water solution were a urine specimen, what would the specific gravity tell you about the condition of the patient?

         _____

         _____

Related Questions for Experiment 1

1. Some brands of pain relievers contain aspirin plus other active ingredients such as acetaminophen, buffers, and caffeine. Does the label of the aspirin bottle used in this experiment list any other active ingredients?

   _____

2. What are the inactive ingredients in aspirin? Hint: Use the *Physician's Desk Reference* [see Contents, Brand and Generic Name Index (Section 2) and/or Product Category Index (Section 3) in the early pages] to look up the "Inactive Ingredients" in any brand of aspirin (give its name) listed.

   Brand and ingredients: _____

   _____

3. What is the purpose of the inactive ingredients in aspirin?

   _____

   _____

4. a. Could you determine the volume of a cube of sugar using the water displacement

      method?   _____

   b. Why or why not?

      _____

Experiment **2**
Some Laboratory Techniques

Reference: *General, Organic, and Biological Chemistry: An Integrated Approach, 4th ed., Chapter 3 and Section 2 of Chapter 5. For this and later Experiments, the Appendix in this manual can also provide helpful information.*

Purpose: To use an exchange-of-partners reaction to familiarize you with filtration and evaporation.

Introduction

LABORATORY HEAT SOURCES

So that you become familiar with the operation and applications of Bunsen burners and hotplates, carefully read Part 5 of Fundamental Laboratory Operations, pages 17-19.

A CHEMICAL REACTION

Many chemicals when placed in water dissolve, forming **aqueous** (water) solutions. When two solutions are mixed, the dissolved chemicals sometimes react. One of the most common types of reaction is the exchange-of-partners type, which can be represented by:

$$A–B + C–D \rightarrow A–D + C–B$$

The partners in the simplest examples are positively and negatively charged species called ions. In the reaction to be studied here, strontium chloride (made up of positive strontium ions and negative chloride ions) reacts with copper sulfate (made up of positive copper ions and negative sulfate ions). By exchanging partners, they form strontium sulfate and copper chloride. When, as here, the exchange of partners results in an insoluble product, that pair of ions precipitates and the other new pair remains dissolved. In such a case it is easy to see that a chemical reaction has occurred.

The situation can be described by writing a word equation:

strontium chloride + copper sulfate → strontium sulfate + copper chloride

We will isolate the insoluble product (the **precipitate**) by gravity filtration and the soluble one by evaporating the water from the **filtrate** (the liquid that runs through the filter).

To identify which of the new chemicals is the insoluble one and which the soluble, we can utilize their properties. The copper ions used here have a +2 ionic charge, so they are properly called copper(II) ions; another type has only a +1 charge and is called copper(I). Copper(II) ions in aqueous solution, or even when just moistened, give their compounds a blue-green color while strontium ions do not make their compounds colored.

Another way to distinguish strontium and copper compounds is by flame test. When placed in a flame or used in fireworks, strontium compounds produce a red color and copper compounds a

Experiment 2   **45**

blue-green color. Some uses of the compounds employed here are: strontium chloride is used as a dental desensitizer; copper(II) sulfate, mixed with calcium hydroxide, is used for spraying grape vines to control fungal infections that are a serious problem for vintners.

Procedure

Use a pen to check off each step as it is completed and for entering information in the Report section. Your instructor may ask for a summary of MSDS information on the chemicals to be used.

CAUTION: Make sure that you wear safety goggles, which is always required when in the laboratory. Especially if using a burner, make sure all loose clothing and long hair is secured/tied back.

THE STUDY OF THE REACTION

I. FILTRATION

1. Review pages 9-15 on the measuring and transferring of liquids. Using a 10-mL graduated cylinder, measure out 5 mL of 0.2 **M**[*] strontium chloride solution and pour it into a 16- x 125-mm test tube. Rinse the graduated cylinder (often just called a "graduate") with deionized water and then use it to measure out 5 mL of 0.2 M copper(II) sulfate solution, which should then be added to the same test tube.

2. Stir with a glass stirring rod and then wait 2 minutes before filtering to allow the precipitate to form and coagulate.

3. Set up equipment for gravity filtration as explained on pages 23-24 and shown in Figures 0-27 and 0-28. Re-stir the contents of the test tube above to get the precipitate into suspension and quickly pour it into the filter paper cone in the funnel. The filtrate (the liquid that comes through the filter) should be clear; if not, re-filter it. [This is a good point at which to make sure that you know that "clear" means free of visible suspended particles. It must be differentiated from "colorless." For example, the filtrate obtained here will not be colorless; it will be pale blue and, if the filtration was effective, should be both clear and blue.] Rinse the precipitate by pouring about a milliliter of deionized water on it and allowing that to be collected with the original filtrate. Keep the precipitate (on the filter paper) for further study; evaporate the water from the filtrate as directed in Part II to isolate the solute.

4. On the Report Sheet, note the colors of the precipitate collected on the filter paper and of the filtrate.

II. EVAPORATION

1. To obtain the solute (dissolved material) that is in the filtrate, pour the filtrate into an evaporating dish, add a boiling stone, and heat gently to boil off the water using one of the procedures explained on pages 21-22 and in Figures 0-24 and 0-25. [A "boiling stone" is a

---

[*] **M** designates a way to specify solution concentration, one that will be studied later.

chip of porous material that helps prevent "bumping," a sometimes dangerous spattering of the hot liquid.]

2. On the Report Sheet, note the color of the residue left in the evaporating dish. If it is completely dry, its color will be neither white nor blue green. To check to see if it really is a copper compound, let the evaporating dish cool and then grind the residue with a couple of drops of water using a stirring rod. If the residue becomes blue green, it is the possible copper compound, copper(II) chloride. Note the result and then from these observations of the colors decide which product is the precipitate and which is the filtrate residue, remembering that one is copper(II) chloride and the other is strontium sulfate.

3. Methods of confirming your conclusions:

    a. Look up the solubilities of the possible products, strontium sulfate and copper(II) chloride, in a reference like the *Handbook of Chemistry and Physics*[*].

    b. If authentic samples of the two compounds are available, test their solubilities in water: Add a few crystals to about 1 mL of water in a test tube and stir.

    c. If a Bunsen burner is available, dip a Nichrome or platinum wire that is held in a pair of forceps or is sealed into a glass tube into the dry compound being tested, place the wire in the flame, and see what color the flame becomes. The wire should be cleaned before each use by dipping it into dilute hydrochloric acid and then heating in the flame. The precipitate may give a stronger flame test if the wire is dipped first into the hydrochloric acid and then into the dry solid just before heating.

CLEANUP

1. Discard the reaction products and any test solutions in the "Waste Salts" container. Any filter paper that contains waste precipitates should be placed in the "Waste Filter Paper" container, not in ordinary wastebaskets. Never discard chemicals into the sinks unless this is specifically recommended.

2. Allow all equipment to cool to about room temperature. Then wash the equipment following the locally specified procedures before putting it away.

3. Wash your hands before leaving the laboratory.

---

[*] Weast, R. C., Ed. CRC Press: Cleveland, Ohio. Virtually any edition will be adequate.

Some Laboratory Techniques

Prelab Questions for Experiment 2

_____ 1. The best way to separate sugar from water is
   a. filtration
   b. evaporation

_____ 2. The best way to separate sand from water is
   a. filtration
   b. evaporation

_____ 3. In a solution of salt in water,
   a. the solute is water and the solvent is salt
   b. the solute is salt and the solvent is water

4. In Part II of this experiment, which are the reactants?

_____ and _____

and which are likely to be the products?

_____ and _____

5. Define *filtrate*.

_____

6. Define *precipitate*.

_____

Report for Experiment 2

I. FILTRATION

1. Color of the precipitate _____

2. Color of the filtrate _____

II. EVAPORATION

1. Color of the residual solid _____

2. Color after moistening _____

3. Conclusions: The precipitate is _____

The soluble product (recovered solute) is _____

4. Briefly note the results of any confirmation tests:

_____

_____

Related Questions for Experiment 2

1. LD_{50} is a measure of toxicity—lethal dose, 50 percent kill. LD_{50} is the dose in mg of the substance which is most likely to cause death within 14 days in one half of the animals used in the testing per kilogram of body weight. The LD_{50} of barium chloride is 118 mg/kg and the LD_{50} of strontium chloride is estimated at 2060 mg/kg. Which compound is more toxic?

_____

2. Barium compounds are chemically very similar to strontium compounds. Water-soluble compounds of both are toxic, but to different degrees. An antidote for barium chloride poisoning is to orally administer 30 grams of sodium sulfate in 250 mL of water. Explain the chemical basis for this antidote.

_____

_____

Experiment 3

Ions; Role in Nutrition

Reference: *General, Organic, and Biological Chemistry: An Integrated Approach, 4th ed., Chapters 2 and 3. See also the Appendix in this manual.*

Purpose: To learn the conductivity method for detecting ions, and to learn methods for identifying some specific ions encountered every day that are vital to living organisms.

Introduction

Experiment 2 introduced ionic compounds and their ability to undergo exchange-of-partners reactions. Additional background for this experiment can be obtained by studying the exercise on chemical periodicity that is in the Appendix.

Of the 90 elements naturally found on earth, 25 are essential to living organisms. The major elements in all living organisms are hydrogen, oxygen, carbon, and nitrogen—they are found in organic compounds such as carbohydrates, fats, proteins, and vitamins. The remaining 21 elements are **minerals**—these are naturally occurring inorganic materials, nearly always ionic. The minerals can be divided into two groups, the major minerals and the trace minerals.

The **major minerals**, or macrominerals, are found in greater mass and include calcium, phosphorus, potassium, sulfur, chloride, sodium, and magnesium. Together, calcium and phosphorus make up three-fourths of the mass of all the minerals present in your body. The **trace minerals** are iron, iodide, fluoride, manganese, zinc, molybdenum, copper, cobalt, chromium, selenium, arsenic, nickel, silicon, and boron. Even though the trace minerals are found in lower amounts than the major minerals they are equally important to your body. A daily deficiency of a few micrograms of iodine is just as serious as a deficiency of several hundred milligrams of calcium. Other minerals such as gold and aluminum are also found in the body but are not known to be nutrients.

The procedures in this experiment are **qualitative** tests—they determine *what* substance is present. These same procedures could be modified for **quantitative** analysis—to determine *how much* of the substance is present. Included are a general test for ions based on electrical conductivity and then specific tests for four ions of nutritional importance. [While two of these ions have positive charges (are cations) and two have negative charges (are anions), it is important to know that in nature they always are accompanied with enough ions of opposite charge so that the combination is electrically neutral.] Iodide ion (colorless) is detected by oxidation to its strongly colored free element form, I_2. The other three of these ions (calcium, phosphate, and iron) are detected by conversion to insoluble products by exchange-of-partners reactions analogous to the one used in Experiment 2.

ELECTROLYTES AND NONELECTROLYTES

When a solute dissolves in a solvent, the solute particles may be charged ionic particles or uncharged molecular particles. Solutions of uncharged particles do not conduct an electric current, and substances that form uncharged molecular particles in aqueous solution are called **nonelectrolytes**. An example is glucose, the sugar found in blood. It exists in water as uncharged molecules so its solutions do not conduct electricity.

$$\text{glucose (s)* } \xrightarrow{\text{ in } H_2O} \text{ glucose (aq)*}$$

Aqueous solutions of **electrolytes** conduct an electric current because they contain charged particles (ions). **Strong** electrolytes in their dry solid crystalline state contain ions that are held in a precise lattice array, but on dissolving these ions become able to move around and then can conduct an electric current. Such ionic compounds usually are water soluble because of hydration of ions, the clustering of polar water molecules around the ions by electrostatic attraction. This stabilizes the dissolved ions, making them less likely to go back into the crystal. An example of an electrolyte is table salt, sodium chloride.

$$\text{NaCl (s)* } \xrightarrow{\text{ in } H_2O} \text{ Na}^+ \text{ (aq) + Cl}^- \text{ (aq)}$$

Weak electrolytes in their water-free state exist in uncharged molecular form, but when dissolved in water these molecules exhibit a limited ability to separate into ions, which makes their solutions slightly conductive. The ions from these weak electrolyte molecules, just as with the ions from a strong electrolyte, are stabilized by the clustering of polar water molecules around them. This encouragement of ionization by adding water to a weak electrolyte will be demonstrated using the weak electrolyte (which as a result is also a weak acid) acetic acid. (Vinegar is a 5% solution of acetic acid in water.)

$$\text{CH}_3\text{COOH (l)* } \rightleftharpoons \xrightarrow{\text{ in } H_2O} \text{ H}^+ \text{ (aq) + CH}_3\text{COO}^- \text{ (aq)}$$

Initially: 100%	~0%
In water: 99%	1%

The conductivity apparatus, which uses a 9-volt battery, tests the electrical conductivity of solutions so their solutes can be classified as strong, weak, or nonelectrolytes. The tester employs a small bulb or one or two diodes (depending on the model) that glow brighter the more conductive the solution into which its electrodes are dipped.

IODIDE ION

The element iodine usually is found in nature as the **iodide ion** (I^-). Iodide ion is needed by the body in an extremely small quantity, but obtaining this amount is critical. The body uses iodide ion to make **thyroxin,** the hormone responsible for regulating the basal metabolic rate. Without iodide ion, the body cannot make thyroxin. Thyroxin is synthesized in the **thyroid,** a gland located in the lower neck. When the iodide level of the blood is low, the cells of the thyroid enlarge, forming a **goiter.** People with this condition suffer sluggishness and weight gain. Infants born to mothers suffering from this condition may be born with irreversible mental and physical retardation known as **cretinism.**

The Food and Nutrition Board, a unit of the National Academy of Sciences, establishes recommendations concerning appropriate nutrient intake for people in the United States. These are the **recommended dietary allowances (RDA**).** RDAs are the dietary intake levels of a nutrient believed to be sufficient to meet the requirements of 97–98% of healthy individuals in specified lifestyle and gender categories. RDAs have not been established for all nutrients but the main RDAs include recommendations for energy (Calories), protein, and numerous vitamins and minerals. Separate RDAs have been set for different sets of people: men, women, pregnant women, lactating women, children, and different age groups. The U.S. RDA of iodine is 0.15 mg for adult men and women.

---

*These are indicators of physical state: **s** means solid, **l** liquid, **g** gas, and **aq** aqueous (that is, dissolved in water).

RDAs are being replaced with **RDIs (Recommended Dietary Intake) aimed at prevention of dietary-related diseases, rather than RDAs, which focused on preventing deficiencies.

Iodide ion is found in some soils and is therefore present in plants growing in that soil. A more plentiful source is the ocean, so seafood is a dependable source of iodine. To prevent dietary shortages of iodine, iodide ion (in the form of potassium iodide) can be added to table salt. The salt-box label will state whether the salt is iodized. Sometimes iodine is also added to milk and bakery products.

Excessive intake of iodine can also enlarge the thyroid. The level at which detectable harm results is only a few times higher than most people receive daily. Like other nutrients, iodide ion is a deadly poison in large amounts. The RDAs attempt to ensure adequate dietary nutrition but they do not warn against overnutrition. Separate recommendations are made for maximum daily intake of nutrients.

In this experiment, salt will be tested for the presence of iodide ion. Iodide ion (I^-) is oxidized* to free iodine (I_2) by chlorine-water, a mixture of Cl_2 and water.

$$2\,I^- \; + \; Cl_2 \longrightarrow 2\,Cl^- \; + \; I_2$$

Iodide ion is very soluble in water; iodine is only slightly soluble in water. Iodine is much more soluble in less polar solvents such as dichloromethane. Dichloromethane and water are **immiscible**—they do not mix. When iodine dissolves in dichloromethane, the solution turns a violet color. This ready extraction of iodine (which is nonpolar since its bond is between like atoms) into the dichloromethane layer illustrates the solubility rule "like dissolves like."

CALCIUM ION

Calcium ion is essential to bone formation, tooth formation, nerve transmission, maintenance of normal blood pressure, blood clotting, muscle contraction, and heart function. Cells need continuous access to calcium, so the body maintains a calcium ion concentration in the blood. The skeleton serves as a bank from which the blood can borrow and return calcium. You can go without adequate calcium for years without suffering noticeable symptoms and then, late in life, discover that your calcium savings are depleted and the integrity of your skeleton can no longer be maintained. Adult bone loss, or **osteoporosis**, is a health problem for many older people. Each year over a million people in the U.S. suffer bone breaks due to osteoporosis. Menopause increases bone loss in women.

The recommended daily allowance of calcium varies by age, ranging between 400 mg for infants and 1200 mg for ages 11-24; the RDA of calcium is also 1200 mg for pregnant and lactating women. After age 24 the RDA is set at 800 mg. The RDA for young people is high because people develop their peak bone mass during this period. Unfortunately, few girls and women meet the RDA during their bone-forming years. This may mean that they start their adult lives with less than optimal bone density.

Calcium ion is not widespread in the diet. Milk and milk products are the most important dietary sources of calcium. People who are unwilling or unable to consume milk products or other calcium-rich foods may choose calcium supplements instead. Regular vitamin-mineral pills do not contain a significant amount of calcium. One problem concerning the use of calcium supplements is possible excessive calcium intake which can cause urinary tract stones, kidney damage, vitamin D toxicity, low iron levels, drug interactions, constipation and confusion.

---

*Elemental iodine (or any other element) has oxidation number 0; iodide ion has oxidation number −1. An **oxidation** is a chemical reaction that increases the oxidation number of a substance. During oxidation a substance loses one or more electrons. A **reduction** is a chemical reaction that decreases the oxidation number of a substance. During reduction a substance gains one or more electrons. Oxidation and reduction must occur simultaneously; while one substance is oxidized another is reduced.

In this experiment, milk will be tested for calcium ion by use of ammonium oxalate, $(NH_4)_2C_2O_4$. If present, a precipitate of calcium oxalate, CaC_2O_4, is formed.

$$(NH_4)_2C_2O_4\,(aq)\ +\ Ca^{2+}\,(aq) \longrightarrow CaC_2O_4\,(s)\ +\ 2\,NH_4^+\,(aq)$$

PHOSPHATE ION

Phosphate ion, PO_4^{3-}, together with calcium ion, is essential to bone and tooth formation. Phosphates are important constituents of **buffers** in the blood which maintain blood acid-base balance. Phosphorus compounds are essential to energy transfer in cells via adenosine triphosphate (ATP). Phosphorus is an essential part of the genetic materials DNA and RNA. Certain lipids—**phospholipids**—that form the membranes around each cell also contain phosphorus.

Phosphorus needs are easily met by most any diet and deficiencies are unknown. The U.S. RDA for phosphorus is 1200 mg for ages 11-24, and for pregnant and lactating women. After age 24 the RDA is 800 mg. Excess phosphorus intake may cause calcium excretion.

In this experiment, milk will be tested for phosphate ion, treating the milk with a few drops each of nitric acid and ammonium molybdate solution. Formation of a precipitate, ammonium phosphomolybdate, confirms the presence of phosphate ion.

IRON ION

Iron ion is primarily needed as a component of **hemoglobin** and **myoglobin**. Both compounds use iron to carry or hold oxygen. Hemoglobin is the oxygen carrier in the blood and myoglobin is the oxygen reservoir in the muscle cells. Iron is also used by many enzymes in the metabolic transfer of energy.

Iron-deficiency anemia is a condition where the red blood cells contain too little iron and thus deliver too little oxygen to the tissues. Symptoms include tiredness, apathy, inability to concentrate, and a tendency to feel cold. In the U.S., some eight percent of women and one percent of men have anemia; many more have only borderline reservoirs of iron. Worldwide, the incidence of iron deficiency is more pronounced—over a billion people suffer from anemia.

Iron deficiency is usually caused by sheer lack of food or high consumption of the wrong foods. In the U.S. the cause is too few iron-rich foods and too many foods high in sugar and fat, which cause poor absorption of iron. Tea, coffee, wheat bran, fiber, and calcium supplements also decrease the absorption of iron. Even a normal person absorbs only about 10 percent of the iron in the diet. Iron in meat, fish, and poultry is more easily absorbed than that in other sources such as egg yolks, legumes, and dried fruits. Vitamin C enhances the absorption of iron.

The iron U.S. RDA is 10 mg for men and older women. For women of childbearing age the RDA is 15 mg; pregnant women should receive 30 mg of iron daily. Of the body's iron, 80 percent is in the blood, so loss of blood causes loss of iron. Because of menstruation, a woman's need for iron is greater than a man's.

The U.S. diet usually does not provide enough iron. It is best to rely on foods to meet this need rather than iron supplements; however, iron supplements are usually recommended for pregnant women. About 25 percent of your iron comes from "enriched" or "fortified" foods that have iron added during processing. These include enriched bread and fortified breakfast cereals. Cooking in iron pans adds substantial amounts of iron to your food just as cooking in aluminum pans adds aluminum to your food. Aluminum is not known to be a nutrient and some have suggested that dietary aluminum may be harmful.

Iron is toxic in large amounts, and once absorbed it is difficult to excrete. Even for healthy men, the safety of high levels of iron intake is questionable. Excess iron may cause liver injury, a greater risk of infections, and possibly increased incidence of heart disease.

Iron can be found in three forms. **Elemental iron** (Fe) is the metallic solid used to make iron pans. Elemental iron atoms are neutral; they have no ionic charge. Tiny iron metal filings are sometimes added to foods. Elemental iron may be listed on an ingredients label as "iron" or as "reduced iron." In the stomach, elemental iron is slowly oxidized* by hydrochloric acid to iron(II) ion.

This second form of iron, iron(II) ion, is also called **ferrous** ion. Iron(II) ion has a positive two charge, Fe^{2+}. Iron(III) ion or **ferric** ion, Fe^{3+}, is the third form of iron. Ferrous iron is more easily absorbed than ferric iron. Ferrous iron is somewhat unstable; it is easily oxidized to ferric ion by the oxygen in the air.

In this experiment, iron in breakfast cereal will be tested for in two ways. Elemental iron is attracted to a magnet; a Teflon-covered magnetic stirring bar can be used to extract the elemental iron from a breakfast cereal.

Iron(II) ion is not attracted to a magnet but can be detected by its reaction with potassium ferricyanide, $K_3Fe(CN)_6$ to form a dark blue precipitate. Iron(III) reacts with potassium ferricyanide to form a green precipitate.

$$3\ Fe^{2+}(aq)\ +\ 2\ Fe(CN)_6^{3-}(aq)\ \longrightarrow\ Fe_3[Fe(CN)_6]_2\,(s)$$
$$\text{(dark blue)}$$

$$Fe^{3+}(aq)\ +\ Fe(CN)_6^{3-}(aq)\ \longrightarrow\ Fe[Fe(CN)_6]\,(s)$$
$$\text{(green)}$$

Procedure

I. ELECTROLYTES

1. Fill separate wells of a porcelein well plate to near maximum capacity with each of the following solutions:

   No. 1: 0.9% salt (NaCl) solution
   No. 2: 1% sugar (sucrose, $C_{12}H_{22}O_{11}$) solution
   No. 3: Vinegar (5% acetic acid, CH_3COOH)
   No. 4: 10% ethanol (C_2H_6O)
   No. 5: Deionized water
   No. 6: 0.1 M** potassium iodide (KI)
   No. 7: 0.1 M trisodium phosphate (Na_3PO_4)
   No. 8: 0.1 M hydrochloric acid (stomach acid, HCl)
   No. 9: 0.1 M sodium hydroxide (lye, NaOH)
   No. 10: 0.1 M sodium bicarbonate (baking soda, $NaHCO_3$)

---

*Elemental iron has an **oxidation number** of zero. The oxidation number of Fe^{2+} is +2 and the oxidation number of Fe^{3+} is +3.

**M is an abbreviation for "molarity," a method of expressing solution concentration (see Raymond, Chapter 6).

2. Dip the tester electrodes successively into each well (rinse them with deionized water between tests), record the brightness of illumination, and classify each substance as a strong electrolyte (S), weak electrolyte (W), or nonelectrolyte (N).

3. Place 5 mL of glacial (100%) acetic acid into a 50-mL beaker and test its conductivity. While monitoring the conductivity, gradually add 10 mL of deionized water to the beaker. Note the results. (**CAUTION: Glacial acetic acid is corrosive and has a very strong odor. Adding water to concentrated acid is generally an unsafe procedure because the enormous amount of heat produced causes spattering, but this is not true for concentrated acetic acid—this particular addition may be done with caution.**)

4. **Disposal:** Pour all test solutions in the "Waste Test Mixtures" container.

II. DETECTION OF IODINE IN IODIZED SALT

1. Using a spatula, select a crystal of iodine, no larger than a grain of rice, and add it to a 16 × 125 mm test tube containing 3 mL of deionized water. Note the color of the solution.

2. Add 1 mL of dichloromethane and stir vigorously, using a glass stirring rod. Note the color of the two layers.

3. Use a mortar and pestle to pulverize approximately 5 g of table salt (it may or may not be iodized). *Record its sample number*. Transfer the salt to a 100-mL beaker.

4. Add 8 to 10 mL of 95% ethanol and place the beaker on a magnetic stirrer. Add a magnetic stirring bar and stir the suspension for approximately 30 minutes. Most of the potassium iodide will dissolve in the ethanol; the sodium chloride will not. (If a magnetic stirrer is not available, stir the solution by hand periodically for about 45 minutes.)

5. Proceed to Part III of this experiment while waiting for the potassium iodide to dissolve. Then follow steps 6 through 11.

6. Using an ordinary funnel, filter the suspension. Collect the filtrate in an evaporating dish. (See Fundamental Laboratory Operations, Part 8, for a discussion of filtration methods.)

7. In a hood, evaporate the solution to dryness over a hot water bath using a hotplate. (See Fundamental Laboratory Operations, Part 7b.)

> **CAUTION: Ethanol is flammable. This procedure should only be done in a fume hood using a hotplate. If the ethanol should catch on fire, cover the evaporating dish with a watch glass using a pair of tongs.**

8. Allow the evaporating dish to cool.

9. Dissolve the residue of the evaporating dish in about 3 mL of deionized water and transfer the solution to a 16 × 125-mm test tube.

10. Add 1 mL of dichloromethane and 3 drops of 6 M hydrochloric acid (HCl) to the test tube. Then add 10 drops of commercial bleach (e.g., Hilex). This will produce the needed chlorine.

11. Stir vigorously as above. Note the color of the two layers.

12. If neither layer is colored, add 5 drops of 0.1 M potassium iodide (KI) solution, stir as before, and note any formation of color.

13. **Disposal:** Pour the dichloromethane solutions into the "Waste Halogenated Solvent" container or as directed by your instructor.

III. DETECTION OF CALCIUM ION IN MILK

1. Pour 6 mL of whole or skim milk into a 100-mL beaker. Record the type used. Add 6 mL of deionized water.

2. Slowly add, while stirring, 3 mL of a 5% solution of acetic acid. The milk fat and casein protein will precipitate.

3. After the precipitate settles, filter the solution using a Buchner funnel, 50-mL filter flask, and trap. Filtration is slow, but only 3 to 4 mL of filtrate are needed.

4. Pour 1 mL of the filtrate into a 10 x 75-mm test tube and add 1 mL of 4% ammonium oxalate, $(NH_4)_2C_2O_4$. A precipitate (calcium oxalate) indicates the presence of calcium.

5. Save the remainder of the filtrate for use in Parts IV and V.

6. Clean the Buchner funnel and filter flask for re-use in Part V.C.

7. Dispose of the contents of the test tubes as in Part I.

IV. DETECTION OF PHOSPHATE ION IN MILK

Follow Steps 1 through 3 with the filtrate from Part III.

1. On a hot plate, heat a 100-mL beaker that is half-filled with water to just short of boiling.

2. Pour 1 mL of the filtrate into a test tube, then add 1 drop of concentrated nitric acid (HNO_3) and 1 drop of 2.5% ammonium molybdate solution.

> **CAUTION: Nitric acid is a strong acid. Wash your hands immediately after handling concentrated nitric acid. Notify your instructor about spills on the bench top.**

3. Place the test tube in the hot-water bath prepared in step 1. The formation of a precipitate indicates the presence of phosphate ion. Record your result.

4. Dispose of the contents of the test tubes as in Part I.

V. DETECTION OF IRON

A. Iron in Milk

1. Pour 1 mL of the milk filtrate from Procedure III into a test tube and add 5 drops of 0.1 M $K_3Fe(CN)_6$ (potassium ferricyanide). A dark blue color indicates the presence of iron(II), green indicates the presence of iron(III).

2. As a method test repeat Step 1 with each of the following:

    (a) 1 mL of 0.1 M iron(II) sulfate solution, and

    (b) 1 mL of 0.1 M iron(III) sulfate solution.

3. Dispose of the contents of the test tubes as in Part I.

B. Elemental Iron in Breakfast Cereal (Optional)

1. Place a very small serving of an instant hot breakfast cereal in a plastic sandwich-size zip-close bag. Add a Teflon-coated magnetic stirring bar to the bag. Securely close the bag.

2. Shake the bag for 2 or 3 minutes.

3. Remove the stirring bar and look for evidence of iron. The iron filings are very tiny. It may help to hold the stirring bar under a gentle stream of water; the water will wash the iron to the end of the stirring bar.

4. Clean the stirring bar by wiping the iron onto a paper towel or by washing off the iron with a strong stream of water.

C. Iron Ions in Breakfast Cereal

Perform Steps 1 to 3 in a fume hood. It is advised that gloves be used to handle the acid and the filtrate.

1. Weigh approximately 2 g of breakfast cereal in a 50-mL beaker, and add 5 mL of 6 M HCl (hydrochloric acid). (**CAUTION!**)

2. Cover the beaker with a small watch glass, and place it for about 15 minutes in a 400-mL beaker half-filled with warm water (70°C).

3. Allow the bath and inner beaker to cool. Then filter the cereal/acid mixture, using the Buchner funnel and filter flask.

4. Pour 1 mL of the filtrate into a test tube and add 5 drops of 0.1 M $K_3Fe(CN)_6$ (potassium ferricyanide). Allow the solution to stand for at least 10 minutes. A dark blue color indicates the presence of iron.

5. Dispose of the contents of the test tubes as in Part I.

Prelab Questions for Experiment 3

_____ 1. What substance turns dichloromethane to a violet color?
  a. iodide ion
  b. iodine
  c. calcium ion
  d. phosphate ion

_____ 2. In Part I, the purpose of adding chlorine water is to
  a. make iodine more soluble.
  b. change iodide ion to iodine.
  c. change iodine to iodide ion.
  d. make dichloromethane immiscible with water.

_____ 3. In Part II, the purpose of adding acetic acid is to
  a. precipitate calcium ion.
  b. remove iodide ion.
  c. precipitate milk fat and casein.
  d. precipitate albumins and globulins.

_____ 4. What substance forms a precipitate with ammonium oxalate?
  a. iodide ion
  b. iodine
  c. calcium ion
  d. phosphate ion

_____ 5. What substance forms a precipitate with ammonium molybdate?
  a. iodide ion
  b. iodine
  c. calcium ion
  d. phosphate ion

_____ 6. What substance reacts with potassium ferricyanide to form a dark blue product?
  a. Fe^{2+} ion
  b. Fe^{3+} ion
  c. both Fe^{2+} and Fe^{3+} ions
  d. Fe

7. What does RDA mean?

_____

_____

8. Draw the electron dot diagram of iodide ion and iodine.

  a. Iodide ion, I^-

  b. Iodine molecule, I_2

Ions of Nutrition

Report for Experiment 3

I. ELECTROLYTES

1. Data

	Solution	Bulb or Diode Brightness (1 = off; 4 = max)	Electrolyte Type (S, W, or N)
1	0.9% NaCl		
2	1% sugar ($C_{12}H_{22}O_{11}$)		
3	Vinegar (5% CH_3COOH)		
4	10% ethanol (C_2H_6O)		
5	Deionized water		
6	0.1 M KI		
7	0.1 M Na_3PO_4		
8	0.1 M HCl		
9	0.1 M NaOH		
10	0.1 M $NaHCO_3$		

Glacial (100%) acetic acid	
Acetic acid diluted with H_2O	

2. Suggest a reason why the conductivities of tapwater and deionized water might be different.

   _____

3. Explain the differing conductivities of 100% acetic acid (CH_3COOH) and diluted acetic acid.

   _____

   _____

II. DETECTION OF IODINE IN IODIZED SALT

1. a. How did the color produced enable you to distinguish the dichloromethane from the water layer?

    b. Is dichloromethane more dense or less dense than water?

2. Record the color of the following solutions.

	Water	Dichloromethane
Iodine (Steps 1 and 2)		
Salt, sample number _____ (Step 11)		
Method test (Step 12)		

3. Was iodide ion present in your salt sample (i.e., was it "iodized")?

    _____

4. If not, why was potassium iodide solution added?

    _____

III. DETECTION OF CALCIUM IONS IN MILK

1. a. What is the difference between whole milk and skim milk?

        _____

    b. Which type did you test? _____

2. Why was acetic acid added?

    _____

3. Describe the evidence for the presence of calcium ion and write an equation for the test reaction.

    _____

4. Oxalic acid reacts with ammonia to form the test reagent used here. One food source of oxalic acid is rhubarb. How could you test it for the presence of oxalic acid?

    _____

IV. DETECTION OF PHOSPHATE ION IN MILK

1. Describe the result of your test.

   _____

2. If a carbonated beverage yields a precipitate in a test like this, what would it mean?

   _____

V. DETECTION OF IRON

A. Iron Ion in Milk

Sample	Results of $K_3Fe(CN)_6$ test	Type of iron ion present
Milk type _____		
$FeSO_4$ solution		
$Fe_2(SO_4)_3$ soln.		

B. Elemental Iron in Breakfast Cereal

Name of Cereal	Result

C. Iron Ions in Breakfast Cereal

Name of Cereal	Result of $K_3Fe(CN)_6$ test	Type of iron ion present

Related Questions for Experiment 3

The Appendix will be of help in answering questions 2 and 3.

1. Dietary sodium occurs as sodium ion, Na⁺. Diets rarely lack sodium and no RDA has been set for sodium. The estimated minimum daily sodium requirement is 500 mg. Excessive intake of sodium has been linked to high blood pressure and heart disease. The recommended maximum daily intake is 2400 mg sodium, the equivalent of about 1 teaspoon of salt. Americans generally consume at least twice that amount. About 10 percent of your sodium intake is due to sodium that occurs naturally in foods. Your kitchen saltshaker may account for as little as 15 percent of your total sodium intake. Up to 75 percent may be due to processed foods. For example, a one half-cup serving of canned chicken noodle soup (less than half a can) contains 950 mg of sodium. Some frequently taken over-the-counter medicines contain surprisingly large amounts of sodium; one dose of Original Alka-Seltzer® with aspirin contains more than 100 mg of sodium.

   Calculate the recommended daily maximum sodium intake in grams.

2. Write the formula (including the charge) of the following ions of nutrition.

   a. calcium ion

   b. chloride ion

   c. magnesium ion

   d. phosphate ion

   e. sodium ion

   f. iron(II) ion (ferric ion)

   g. iron(III) ion (ferric ion)

   h. fluoride ion

   i. iodide ion

   j. sulfate ion

3. Write the formula of the following compounds found as food additives or food preservatives.

a. potassium iodide

b. sodium bicarbonate

c. sulfur dioxide

d. calcium carbonate

e. sodium chloride

f. ferrous sulfate

g. silicon dioxide

h. ammonium sulfate

i. magnesium carbonate

j. ferric phosphate

k. sodium sulfite

l. sodium bisulfite

m. potassium chloride

n. magnesium oxide

o. sodium carbonate

p. calcium sulfate

q. sodium dihydrogen phosphate

Experiment **4**

Paper and Thin-Layer Chromatography

Purpose: To become familiar with two chromatographic methods of analysis by identifying household air contaminents and by separating the components of food coloring and the pigments found in spinach.

Introduction

Most samples of matter are impure mixtures of two or more substances. Separating and identifying the components of a mixture are important goals in chemistry. One common and powerful separation technique is **chromatography**. The word chromatography means "separation of colors" but today chromatography is used for both colored and colorless substances. The separation process is based on the fact that a porous solid **adsorbs*** different substances to different extremes. The components of the mixture become separated between the stationary porous solid phase and the mobile liquid phase, which passes through the stationary phase. As a liquid mobile phase moves by capillary action, it "pulls" along various molecules that have been placed on the solid stationary phase. Different types of molecules have different attractions to the mobile and to the stationary phases, and therefore do not travel at the same speed through the stationary phase. This leads to a separation of the various molecules.

The simplest types of chromatography, paper and thin-layer (TLC), will be used in this experiment. Other chromatographic methods, including column chromatography, gas chromatography (GC), and high-performance liquid chromatography (HPLC) are used extensively in chemistry and related fields such as medicine.

In medicine, chromatography is used to separate and identify amino acids and proteins in mixtures. Chromatograms of blood samples will sometimes reveal the presence of foreign proteins associated with certain diseases. Law enforcement agencies sometimes require chromatographic analysis of urine specimens from suspected drug addicts.

PAPER CHROMATOGRAPHY

In paper chromatography the stationary phase is a sheet of absorbent paper, such as filter paper. A tiny drop of the mixture to be separated is placed on the paper near the bottom of the paper. A lightly drawn pencil line marks the location of the spot. This location is called the **origin**. The paper is suspended vertically in the mobile phase, a solvent or **eluent**. The eluent could be water, another pure liquid such as alcohol, or a solution. Different eluents will differ in their ability to "pull" along the substances in the mixture being separated.

***Adsorption** is the adhesion of a substance to the *surface* of another substance, as opposed to **absorption**, in which a substance penetrates into the inner structure of another substance.

The origin must be above the surface of the eluent. If not, the applied samples get dissolved off the paper. The eluent rises up the paper by capillary action. When the eluent reaches the origin, the components of the mixture rise at different rates. The container must be covered to prevent evaporation of eluent. The chromatogram must be removed from the eluent before the eluent reaches the top of the paper.

As the substances in the mixture rise up the paper, they spread out and the spots become larger. For this reason, the original spot should be as small as possible, less than 3 or 4 mm in diameter. If too much material is applied to the small spot or the sample is too concentrated, the spot may develop a long "tail." See Figure 4-1. If too little material is applied to the spot, the color of the spot may be too faint to see as the spot enlarges while moving up the paper. Trial-and-error and experience help the experimenter obtain both a small spot and one with the proper amount of material.

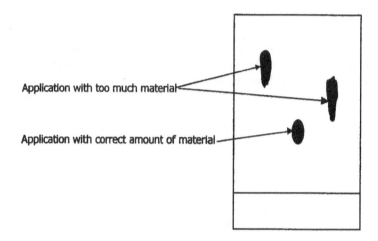

Figure 4-1 Chromatogram "tailing."

A substance can be characterized by its R_f (rate of flow) value, defined as the ratio of:

$$\frac{\text{Distance traveled by the substance from the origin}}{\text{Distance traveled by the eluent from the origin}}$$

An R_f value is a constant for that substance on the same chromatographic sheet but will vary slightly from one chromatogram to another because of variation in such things as eluent composition, moisture content of the chromatographic sheet, or temperature. So it is always best to run an authentic sample on the same sheet.

Figure 4-2 shows the finished chromatogram of substance A, substance B, and a mixture containing substances A and B. For each substance, measure the distance from the origin to the center of the migrated spot. If the spot is large with a "tail," measure to the "center of gravity" of the spot.

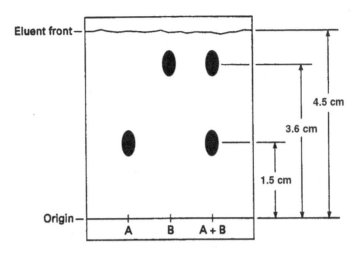

Figure 4-2 Typical finished chromatogram of two substances.

$$R_f \text{ (substance A)} = \frac{1.5 \text{ cm}}{4.5 \text{ cm}} = 0.33$$

$$R_f \text{ (substance B)} = \frac{3.6 \text{ cm}}{4.5 \text{ cm}} = 0.80$$

Once the R_f value is known, the substance can sometimes be identified by comparing its R_f value with those reported in the literature. To check the identity of an unknown substance, it is usually necessary to run a chromatogram of a known sample simultaneously with the unknown.

THIN-LAYER CHROMATOGRAPHY (TLC)

Thin-layer chromatography is almost identical to paper chromatography. Instead of using paper, the stationary phase is a thin coating of adsorbent material, called the **sorbent**, on a sheet of glass, plastic, or metal. As in paper chromatography, the TLC sheet is suspended vertically in an eluent and the eluent travels up the sheet. TLC offers two advantages over paper chromatography. First, it provides a better separation of the mixture with less spreading of the spots; second, the sorbent may be varied. Common sorbents include silica (SiO_2), alumina (Al_2O_3), and cellulose. In order to separate the substances in a mixture, the substances must have different R_f values. By carefully choosing an eluent and a sorbent, it is possible to find a combination that will separate almost any mixture.

FOOD COLORING

Color is important in consumer acceptance of a product. Even the color of the container can make a difference in consumer purchases. For these reasons people have been coloring foods, drugs, and cosmetic products for thousands of years. Food was once colored with dyes such as those from beets, peppers, grape skins, saffron, and insect bodies. In the 19th century as the country changed from an agricultural to an industrialized nation, colorants containing lead, copper, arsenic, or other toxic substances became more widely used and presented significant health risks. In response, the Federal Food and Drug Act of 1906 and the Federal Food, Drug and Cosmetic Act of 1938 required government certification and control over color additives for drugs and cosmetics as well as foods. A 1960 amendment to the FD&C Act included the Delaney Clause which banned additives shown to induce cancer in humans or animals, even at very low doses. The Delaney Clause was based on the premise that there is no safe threshold for cancer-causing substances.

The U.S. Food and Drug Administration banned Red Dye No. 2 in 1976 because it was suspected, but not proven, to cause cancer. The FDA proposed banning the artificial sweetener saccharine in 1977 because studies showed it could cause cancer. Public outcry over the loss of this sweetener forced Congress to specifically exempt saccharine from the Delaney Clause. Interestingly, based on the same research studies, Canada prohibits most uses of saccharine and permits use of the sweetener cyclamate and Red Dye No. 2, exactly the opposite of the United States.

One of the seven FD&C approved food colorants, Red No. 3, has been shown to cause thyroid cancer in male rats. Food manufacturers and the cosmetic industry fought off more than a dozen attempts to prohibit Red No. 3. Finally the FDA banned one form of Red Dye No. 3; Red No. 3 Lake pigments previously used in products such as lipsticks, candies, and pill coatings were banned in 1990. Lake pigments are insoluble versions of the colorant. Unlike a lake, a dye is soluble in water. The dye version of Red 3 is still permitted in many drugs and foods such as baked goods, dairy products, desserts, jellies, vegetable products, gelatin desserts, pistachio nuts, and powdered beverage mixes. One clearly visible use is in maraschino cherries. Red No. 3 is the only dye that doesn't bleed in citrus juices. Without it, a fruit-cocktail cherry would be brown.

Many have argued that the risk of human cancer from Red No. 3 is extremely small. The Delaney Clause requires the FDA to ban additives if there is any cancer risk, no matter how small. Some have suggested that the Delaney Clause be replaced by a standard that allows for a small, insignificant amount of cancer risk, such as one cancer case in 1 million people. Others have argued that while there are some cancer risks that are unavoidable, we ought to eliminate all the avoidable risks. Food colorants are clearly avoidable because they are intentionally added. If the zero-risk Delaney Clause remains in force, the FDA is someday likely to totally ban Red 3.

Seven "certified" synthetic FD&C dyes can be added to food products: Red 3, Red 40, Blue 1, Blue 2, Yellow 6, and Green 3. Since Yellow 5 causes severe allergic reactions in some people (about 1 in 10,000), the FDA requires its listing on food labels. Yellow 5 was once blamed for hyperactivity in some children but now it is clear that most hyperactivity is not caused by this or other additives. The other six food dyes do not have to be separately identified on food labels and can be listed as "artificial colors." The average U.S. citizen annually consumes about 3 grams of these dyes. Twenty-six other color additives are exempt from certification and are permitted in food—these include caramel, grape skin extract, paprika, turmeric, iron extract, titanium dioxide, and β-carotene. Most exempt colorants are derived from vegetable or mineral sources and are therefore often called "natural"; they can also be made synthetically.

Out of 200 colorants once approved in the United States, only these seven certified and 26 exempt colorants remain approved for addition to our food.

This experiment will examine food colors and determine why grape-flavored Kool-Aid® has a purple color.

DETECTION OF HOUSEHOLD AIR CONTAMINANTS

The object of this experiment is to illustrate the ability of thin layer chromatography (TLC) to (a) determine the presence of possible compounds, (b) distinguish between similar compounds, and (c) show the importance of using known reference compounds.

TLC can separate components of a mixture or provide evidence that a chemical sample is homogeneous. Apparent homogeneity indicated by one form of chromatography, however, may be the result of two or more mixture components behaving identically, so apparent homogeneity always needs to be confirmed by other methods.

This experiment will attempt to identify the presence or absence of two carbonyl compounds, one an aldehyde (formaldehyde) and one a ketone (2-butanone, also called methyl ethyl ketone or MEK). They may both be found in household air. For example, formaldehyde can be released from such sources as glues used to hold together the layers of wood in plywood and 2-butanone from nail polish and other lacquers.

$$R-\overset{\overset{\displaystyle O}{\|}}{C}-H \qquad H-\overset{\overset{\displaystyle O}{\|}}{C}-H \qquad R-\overset{\overset{\displaystyle O}{\|}}{C}-R \qquad CH_3-\overset{\overset{\displaystyle O}{\|}}{C}-CH_2CH_3$$

| aldehyde general structure | formaldehyde | ketone general structure | 2-butanone |

[R represents any of a variety of "groups" that contain carbon and hydrogen atoms.]

Because these compounds are very volatile and easily oxidized (in the case of aldehydes), they can advantageously be converted to non-volatile, stable, and, of special advantage here, colored derivatives by reaction with 2,4-dinitrophenylhydrazine. The equation for the reaction is:

$$R-\overset{O}{\underset{R}{C}} + H_2NNH-\overset{NO_2}{\underset{O_2N}{\bigcirc}} \rightarrow R-\overset{NNH}{\underset{R}{C}}-\overset{NO_2}{\underset{O_2N}{\bigcirc}} + H_2O$$

You will be provided with solutions (in ethyl acetate) of the dinitrophenylhydrazine derivatives (dinitrophenylhydrazones, DNP's) of these two carbonyl compounds that could have been obtained from household air samples. For example, the above equation illustrates formation of the DNP derivative of formaldehyde when both R's = H, or of 2-butanone when one R = CH_3 (a methyl group) and the other CH_3CH_2 (an ethyl group).

For each spot that you see on your developed TLC plates, report the R_f value (the ratio of distance traveled by the spot to distance traveled by the eluent, which here is toluene). As mentioned above, R_f values are somewhat variable from one plate to another because of such things as the degree of dryness of the plates. However, they are constant for a given plate, so if the R_f of an unknown matches that of an authentic sample run on the same plate it is convincing evidence of identity.

SEPARATION OF SPINACH PIGMENTS

Deeply colored vegetables such as spinach contain a mixture of pigments including chlorophyll-a (blue-green), chlorophyll-b (yellow-green), and β-carotene (yellow-orange). When exposed to air the chlorophyll pigments are slowly oxidized to form brown-colored products. The pigments are nonpolar and do not dissolve in water, a highly polar solvent; that's why grass stains are so difficult to launder from clothing. The pigments do dissolve in acetone, a common solvent found in fingernail polish remover. The eluent for the chromatography of these pigments will be a 2:1 mixture of ligroin and acetone. Ligroin is a non-polar solvent similar to gasoline, mineral spirits, or painter's naphtha—it is a mixture of hydrocarbons with a boiling point range of 130–145°C. The 2:1 ligroin-acetone eluent mixture is chosen because its polarity gives a good separation of the spinach pigments. The eluent mixture must be free of water—one drop of water would considerably change the polarity of the mixture.

Procedure

> **CAUTION: Acetone, ligroin, and toluene are flammable. Do not use near open flames.**
>
> **Pregnant women may be advised by their physicians to avoid breathing the vapors of such solvents.**

I. PAPER CHROMATOGRAPHY: FOOD COLORING

1. Cut a 46- × 56-cm piece of chromatography paper into four strips, each 14- × 46-cm. Cut a 9- × 14-cm piece from one of the strips.

2. Use a pencil to draw the origin line about 1.5 cm from the bottom of the 14-cm edge of the paper. Use a pencil to make marks every 1.5 cm along the length of the origin line, starting about 2.5 cm from the edge of the paper. See Figure 4-3.

3. Place 1 drop of food color on a watch glass or in a small beaker. Soak a wooden toothpick in the food color for several seconds then briefly touch the toothpick to the first mark.

4. Use a pencil to write the name of the food color *below* the origin line.

5. Repeat Steps 3 and 4 with each of the other food colors. Use a fresh toothpick for each color.

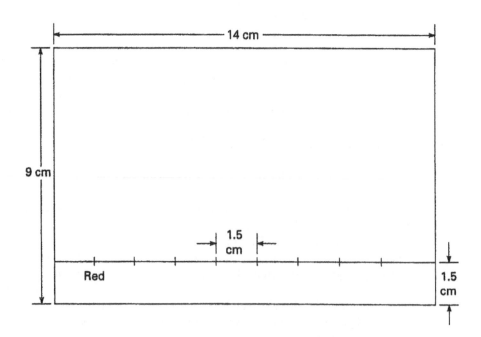

Figure 4-3 Labeling the origin line on the chromatography paper.

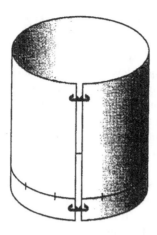

Figure 4-4 Roll into a cylinder and staple with a gap between the ends of the paper.

6. Dissolve 1 packet (3.9 g) of unsweetened grape-flavored Kool-Aid® in 5 mL of water. Stir with a glass stirring rod. Use a fresh toothpick to spot the grape Kool-Aid to the chromatography paper. Reapply the Kool-Aid to the same spot 5 or 6 times to get a darker spot. Use a pencil to write the name "grape" *below* the origin line.

7. Roll the paper into a cylinder with the origin line at the bottom and the dye spots on the outside of the cylinder. Use two fingers of one hand to hold the ends of the paper close together, about 2 mm apart. Staple the top ends and then staple the bottom ends. The ends of the paper should not touch. See Figure 4-4.

8. Add about 10 mL of 0.1% salt (NaCl) solution to a 400-mL beaker. The solution should be about 0.5-cm deep.

9. Carefully place the paper cylinder into the beaker containing the salt solution. Make sure the origin line is at the bottom. The origin line must not be below the surface of the eluent. The paper should not touch the side of the beaker.

10. Cover the beaker with a watch glass. Do not disturb the beaker while the eluent rises up the paper.

11. Watch the eluent as it moves up the paper and see what happens as it comes into contact with the food colors. Leave the paper in the beaker until the eluent front is about 1.5 cm from the top of the paper.

12. When the eluent front nears the top, remove the chromatogram and open the cylinder by tearing the paper at the staples. Set the chromatogram on an empty beaker or a paper towel to dry. Because of the large amount of water absorbed by the chromatography paper, the eluent front may continue to move up the paper for several minutes, dragging the dyes with it. After 2 or 3 minutes, or when the paper appears to be drying out, mark the final position of the eluent front with a pencil line.

13. Does each color contain 1 dye or a mixture of dyes? Calculate the R_f value of each dye. Try to identify the name (Red 40, etc.) of each dye. Identify the dyes present in the grape Kool-Aid.

14. **Disposal:** These paper chromatograms may be discarded in any wastebasket.

II. TLC: HOUSEHOLD AIR CONTAMINANTS

1. Prepare a TLC developing chamber from a 400-mL beaker by adding just enough toluene to bring the liquid level to a level of about 1 cm (safely short of the level at which the sample spots will be placed on the TLC sheet; see below). As a wick, set a filter paper strip (about 3 x 8 cm) into the toluene so that it touches the inside of the beaker. Then tightly cover the beaker with aluminum foil and let this developing chamber sit so its liquid and vapor spaces can equilibrate while you prepare the TLC sheet (see below). *Label your beaker and keep it in a hood to minimize exposure to toluene vapor.*

2. Obtain a 4- x 10-cm silica gel sheet; it must be handled gently and only by the edges (or by the end opposite that near which the samples will be applied). Using a pencil, at 1.5 cm from the narrow edge (the bottom) of the sheet make a light dot on each edge (*only on the edges*; don't draw a line across as that may cause the silica gel to flake off). This will indicate the origin line on which test spots are to be applied. Place three very light dots along this hypothetical line, 1 cm from the edges and 1 cm apart.

3. Using the DNP solutions and 10-μL micropipets supplied, apply 3- to 4-μL portions (this is the volume that is drawn into the pipets by capillary action) of each of the solutions on the plate as directed below by lightly touching the pipet tip to the appropriate spot on the TLC plate. The entirety of each 3- to 4-μL *sample should not be applied all at once*; make the contact with the TLC plate very brief so that the spots are as small in diameter as possible. This should require about three contacts.

   Left lane: Solution of formaldehyde DNP
   Middle lane: Solution containing unknown DNP's (be sure to record the *unknown number* in your manual on the report sheet)
   Right lane: Solution of 2-butanone DNP

4. Then place the TLC sheet in the developing chamber, being careful that it does not dip so deeply that the applied spots get immersed. Lean the sheet so that only the bottom edge and the top corners touch the beaker surface. Put the foil back on and leave the tank until the toluene has crept by capillary action up to about 1 cm from the top of the sheet. Then remove the foil and *immediately* mark the solvent front (before it evaporates) with a pencil.

5. After the sheet has dried (in the hood), place a small pencil dot **in the center** of each spot, measure *to the nearest millimeter* the distances traveled by the solvent front and by each spot, and calculate the R_f values. *(If your chromatogram shows streaks instead of discrete spots, ask your instructor how to proceed.)* Use the table provided in the Report section for recording the chromatography results; add more lines if you see more than the anticipated spots.

6. **Disposal:** Place excess eluent in the "Waste Organic Solvent" container and the TLC sheets in the designated container.

III. TLC: SEPARATION OF SPINACH PIGMENTS

CAUTION: If possible, perform all parts of this procedure that use acetone and the acetone-ligroin mixture in a fume hood.

1. Use a mortar and pestle to crush about 4 grams (about a tablespoon) of spinach into very small pieces. The crushing helps break cell walls and free the pigments from the cells.

2. Place the spinach in a clean, dry 150-mL beaker and add 10 mL of acetone. **CAUTION: Acetone is flammable. No flames present!** Stir with a glass stirring rod for about 10 minutes.

3. Allow the sediment to settle to the bottom of the beaker.

4. Using a pencil, mark the origin on a 1.5- × 10-cm silica gel TLC sheet. Mark the origin lightly at each edge of the silica gel sheet about 1.5 cm from the bottom. Do not use a pen. Do not draw a line completely across the TLC sheet since the sorbent may flake off.

5. Dip a clean capillary tube into the spinach extract. Midway between the 1.5-cm marks, apply the spinach extract to the origin of the silica gel TLC sheet by quickly touching the capillary tube to the sheet. Hold the capillary tube at right angles to the sheet. Do not scrape off the sorbent with the capillary tube. Allow the spot to dry for 30 seconds and repeat the application about 15 times. The drying period is necessary to ensure a small spot. The repeated application is necessary to ensure sufficient material is applied.

6. Pour 2 mL of the 2:1 ligroin-acetone eluent mixture into a dry 25- × 150-mm test tube. The 10-mL graduated cylinder used to measure the ligroin-acetone mixture must be dry—we don't want water added to the mixture. **CAUTION: Ligroin and acetone are flammable. Keep away from flames.** Stand the test tube in a 250-mL Erlenmeyer flask.

7. Lower the TLC sheet into the test tube and stopper the test tube. The origin on the TLC sheet must be above the surface of the ligroin-acetone eluent mixture. Do not disturb the test tube until the eluent front is 1 cm from the top of the TLC sheet. It may take 40–50 minutes for the eluent front to reach that point.

8. Rinse the graduated cylinder with a small amount of acetone and pour into the "Waste Organic Solvent" container. Wash the graduated cylinder with a brush and laboratory soap solution or the detergent designated by your instructor; rinsing with water will not remove non-polar substances such as ligroin.

9. Use forceps to remove the TLC sheet when the eluent is 1 cm from the top of the TLC sheet. Do not allow the eluent to reach the top of the sheet. Immediately mark the position of the eluent front. This eluent evaporates very rapidly and soon you will no longer be able to see the position of the eluent front.

10. Allow the sheet to dry and circle all visible spots. The spots may fade or change colors after exposure to air and light.

11. Beginning at the origin, label all spots as A, B, C, D, etc. and calculate their R_f values.

12. Try to identify the pigments by their colors. Chlorophyll-a is blue-green, chlorophyll-b is yellow-green, and β-carotene is yellow-orange.

13. **Disposal:** Place the excess ligroin-acetone eluent mixture and the excess acetone-spinach extract in the "Waste Organic Solvent" container. Place the excess spinach in the "Waste Spinach" container and the paper chromatograms in any wastebasket.

Prelab Questions for Experiment 4

1. What safety precautions are necessary when using acetone, ligroin, and toluene?

   <u>use gloves, goggles, hazardous waste</u>

2. Define the following terms:

   a. Sorbent _____

   b. Eluent _____

   c. Eluent front _____

   d. Origin _____

3. Why is a pencil used to mark the origin line and not a ball-point or ink pen?

   _____

   _____

4. List three errors that would prevent an accurate R_f determination in the TLC setup illustrated in Figure 4-5.

   a. _____

   b. _____

   c. _____

Front view

Figure 4-5 A student TLC setup with errors.

5. Calculate the R_f value for each substance on the chromatogram in Figure 4-6.

    a. R_f of substance A   _____

        Calculations:

    b. R_f of substance B   _____

        Calculations:

Figure 4-6 A finished chromatogram.

Name _____ Section _____ Date _____

Report for Experiment 4

I. PAPER CHROMATOGRAPHY: FOOD COLORING

Yellow Food Color   Eluent distance   _____

Color of Spot	Distance Spot Traveled	R_f	Identity of Dye
_____	_____	_____	_____
_____	_____	_____	_____

Calculations:

Red Food Color   Eluent distance   _____

Color of Spot	Distance Spot Traveled	R_f	Identity of Dye
_____	_____	_____	_____
_____	_____	_____	_____

Calculations:

Blue Food Color   Eluent distance   _____

Color of Spot	Distance Spot Traveled	R_f	Identity of Dye
_____	_____	_____	_____
_____	_____	_____	_____

Calculations:

Green Food Color   Eluent distance   _____

Color of Spot	Distance Spot Traveled	R_f	Identity of Dye
_____	_____	_____	_____
_____	_____	_____	_____

Calculations:

Orange Food Color   Eluent distance   _____

Color of Spot	Distance Spot Traveled	R_f	Identity of Dye
_____	_____	_____	_____
_____	_____	_____	_____

Calculations:

Grape-flavored Kool-Aid®   Eluent distance   _____

Color of Spot	Distance Spot Traveled	R_f	Identity of Dye
_____	_____	_____	_____
_____	_____	_____	_____

Calculations:

Draw your paper chromatogram here.

II. TLC: HOUSEHOLD AIR CONTAMINANTS

UNKNOWN NUMBER _____

Eluent Distance _____

Lane	Color of Spot	Distance Spot Traveled	R_f	Identity of Spot
Left				
Center				
Right				

Enter data here for any unexpected spots that may be seen:

Answer these questions:

1. Did the procedure distinguish between formaldehyde and 2-butanone DNP's (i.e., are their R_f values distinguishably different)?

2. Which, if any, DNP's did your unknown (**be sure its number is given above**) contain?

3. Compare your R_f values with those of two other pairs of experimenters. By what percentage do your results for the two DNP's differ from the average? *SHOW your data and calculations.*

4. Draw your TLC chromatogram here:

III. TLC: SEPARATION OF SPINACH PIGMENTS

Eluent Distance _____

Spot	Color of Spot	Distance Spot Traveled	R_f	Possible Identity of Pigment
A	_____	_____	_____	_____
B	_____	_____	_____	_____
C	_____	_____	_____	_____
D	_____	_____	_____	_____
E	_____	_____	_____	_____
F	_____	_____	_____	_____

Calculations:

Draw your TLC chromatogram here.

Related Questions for Experiment 4

1. The eluent for paper chromatography of food colors was 0.1% NaCl which gives a better separation of dyes than does deionized water. In addition to artificial colors, the spot of grape Kool-Aid contained citric acid, calcium phosphate, salt, maltodextrin, cornstarch, artificial flavor, and ascorbic acid. Since these substances dissolve in water, the eluent moving up the paper was somewhat different for the Kool-Aid than for the food colors. Could this difference have affected the R_f values of the dyes? Is there any evidence to support your answer?

   _____

   _____

2. Compound X has an R_f value of 0.25.

   a. How far will compound X have traveled from the origin when the eluent front has traveled 4 cm?

      Calculations: _____

   b. How far will compound X have traveled from the origin when the eluent front has traveled 8 cm?

      Calculations: _____

3. Why can an R_f value never be greater than 1.00?

   _____

   _____

4. Carrot pigments were separated by TLC chromatography. The chromatogram was removed when the eluent front was only half-way up the TLC sheet.

   a. What is the advantage of allowing the eluent front to rise to near the top of the TLC sheet rather than stopping when only half-way up?

   _____

   _____

   b. Would the R_f values of the pigments differ if the eluent front rose only half-way rather than to near the top of the TLC sheet?

   _____

   _____

5. A student attempted to separate tomato pigments using this procedure—tomato extract was spotted on a silica gel TLC sheet, developed in a 2:1 ligroin:acetone eluent, and removed when the eluent neared the top of the sheet. Unfortunately, two pigments had the same R_f value and were not separated. What conditions would you change in an effort to obtain satisfactory results?

   _____

   _____

6. Look up the entry for FD&C Yellow No. 5 in *The Merck Index*.

   a. What is the common name for FD&C Yellow No. 5? (This is the first name listed.)

   _____

   b. What is the complete chemical name for FD&C Yellow No. 5? (This is the second name listed—this name describes the structure of Yellow No. 5.)

   _____

   _____

   c. What is the molecular formula of Yellow No. 5?   _____

   d. What is the molecular weight (mol wt) of Yellow No. 5?   _____

   e. What uses are listed?

   _____

   _____

Molecules; Organic Compounds

Reference: *General, Organic and Biological Chemistry: An Integrated Approach, 4th ed., Chapter 4.*

Purpose: You will learn about covalent bonds, bond polarity, and the classes of organic compounds, including their three-dimensional nature.

Introduction

Molecules are groups of atoms held together by the sharing of electrons in what are called covalent bonds, each of which involves the sharing of pairs of electrons. Single bonds have one pair of shared electrons, double bonds have two, and triple bonds have three. If the bonded atoms are identical [as in iodine (I_2)] or differ in electronegativity* by less than 0.5 [as in methane (CH_4)], the bonds are said to be non-polar; if they differ in electronegativity by 0.5 to 1.7 [as in water (H_2O)], the bonds are polar covalent; if by more than that, electrons are not shared and the bonds are ionic [as in non-molecular, ionic sodium chloride (NaCl)]. Here it will be non-polar and polar molecular compounds having skeletons composed only or predominantly of carbon atoms that will be studied. They are called "organic" because the earliest known ones were derived from plants or animals, but since the mid 1800s it has been known that they can be synthesized from lifeless material. They are exceedingly numerous because carbon has limitless ability to form bonds with itself.

If a compound contains only carbon and hydrogen (the element that is almost always accompanying carbon), it is called a **hydrocarbon.** If one or more "hetero" atoms are present, it makes the compound a member of one of the "classes" of hydrocarbon derivatives. Examples of these are the alcohols and carboxylic acids, whose hetero atom-containing "functional groups" are respectively the hydroxyl group (OH) and carboxyl group (COOH). These and other classes of organic compounds will be studied in more depth in later experiments. The emphasis here will be on the two broad classes of hydrocarbons, the **aliphatic** and **aromatic** compounds.

A very important point to remember in beginning the study of organic compounds is that carbon atoms always share four pairs of electrons, two in each covalent bond. It may be in four single bonds, two single bonds and one double bond, or one single bond and one triple bond.

ALIPHATIC HYDROCARBONS

These are further divided into alkanes, alkenes, and alkynes. The great majority of hydrocarbons are colorless, but alkenes that have eight or more adjacent double bonds alternating with single bonds are colored (this kind of alternate bonding is termed "conjugated"). An example is carotene,

---

*The Appendix presents an explanation of electronegativity. See also Raymond, 4th ed., Chapter 4.

a pigment found in carrots and other vegetables. Its structure is shown here in the "skeletal" method, explained below.

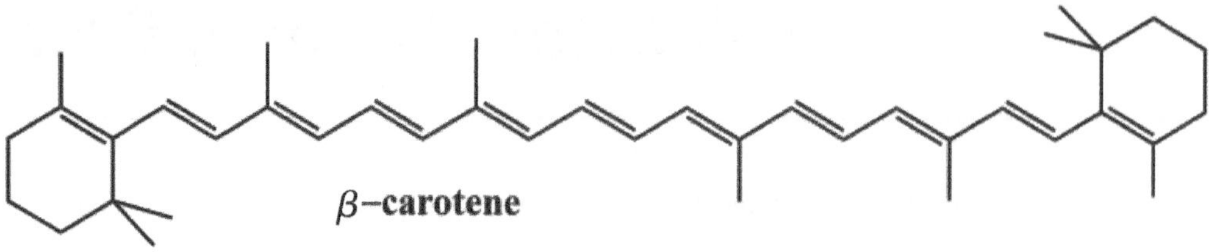

β-carotene

Alkanes

These are the most fundamental type of hydrocarbons and involve only single bonds, four per carbon atom. They contain from one to hundreds of carbon atoms in chains that may be unbranched or branched. The names of the first four, listed in order of increasing numbers of carbon atoms per molecule, are methane, ethane, propane, and butane. With five or more carbons, they are given systematic names: pentane, hexane, heptane, octane, nonane, decane, and so forth. The names of the first four are so frequently needed that they have to be memorized; the others can usually be recalled when needed. The structure of methane, shown in various manners, is:

Methane:

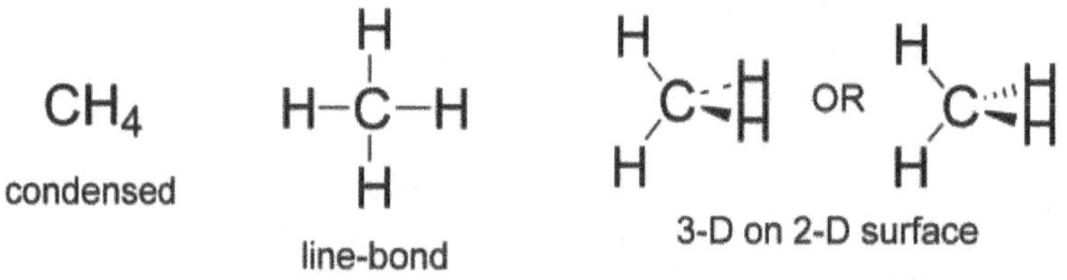

The equivalent versions on the right show the two hydrogen atoms drawn on their left as being in the plane of the paper and the two on the right as being one above the plane (by use of the solid wedge bond) and the other below the plane (either by use of the broken line bond or the dashed wedge bond, equally acceptable methods). This 3-D method is a way of indicating that four bonds on carbon point to the corners of a tetrahedron and are all at 109.5° angles, not the 90° angles that the planar line-bond drawing seemingly implies. The use of molecular models in this experiment will better clarify this.

Propane has a three-carbon chain and provides opportunity to realize that, because of the tetrahedral nature of alkane carbons, it has a zig-zag skeleton.

Propane:

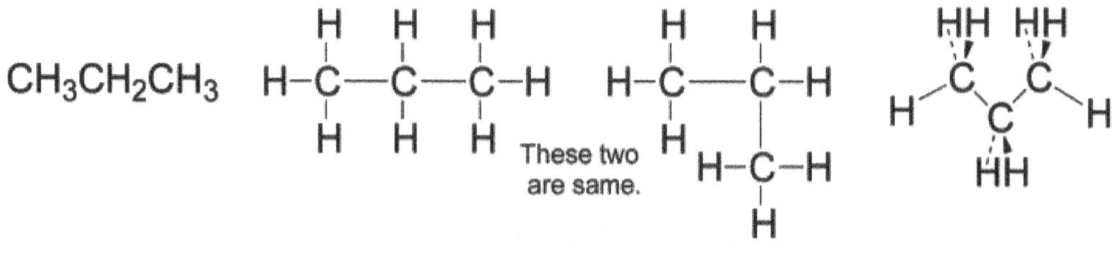

All four drawings are correct. The second and third are equivalent because (1) C-C single bonds allow rotation, and (2) if the bond angles are drawn correctly (as in the far right drawing), they are all tetrahedral 109.5° angles and the chain is always bent at each C-C bond. Again, the exercises using molecular models will clarify this.

With the next longer alkane there is the first opportunity to have either an unbranched chain or a branched chain. These two compounds, called "butane" and "isobutane", respectively, are examples of **constitutional isomers**, compounds that have the same molecular formula but different structures.

Butane isomers:

butane

isobutane
(2-methylpropane)

The names "methane" through "butane" are examples of what are called "common names," ones that came into being before systematic naming was found necessary. The alternative name for "isobutane," 2-methylpropane, is an example of a systematic name. It means that a methyl group (one derived from methane) has been introduced at the second carbon of the propane chain by replacement of a hydrogen atom. Groups like methyl are formed by loss of a hydrogen atom from a parent molecule and named by changing the parent name so that it ends in –yl. So there is ethyl from ethane, propyl from propane, etc. A capital **R** is often used to represent an unspecified alkyl group.

The rotation allowed by single bonds allows a molecule to take on (and move between) different shapes called **conformations**, all of which have the same structure but different angular relationships between the groups at the ends of a single bond. A helpful way of representing these alternative shapes is to use Newman projection drawings. These show the view of a molecule as would be seen when sighting down the bond that connects two atoms. Examples are these different conformations for butane that show the angular relationships between the methyl (CH_3) groups at each end of the molecule when sighting down butane's central C-C bond:

staggered
(most favored)

skew

nearly
eclipsed

In these representations the back carbon atom (represented by the circle) is rotated relative to the front one. The two conformations on the left are the best and second best in terms of stability; the representation on the right shows rotation almost to the eclipsed conformation in which all bonds are aligned. This is especially destabilizing in terms of the alignment of the methyl groups, which are much larger than the hydrogen atoms held by the other bonds.

Alkenes

The alkenes are hydrocarbons that contain one or more double bonds between carbon atoms. This involves the loss of two hydrogen atoms per double bond. Consequently, alkenes (and alkynes) are said to be **unsaturated hydrocarbons** since hydrogen, bromine, or other atoms can be added to their multiple bonds. Note the small but critical change in name suffix that distinguishes alkenes from alkanes. The alkene double bond is the first of the functional groups studied here.

Alkenes:

ethene   propene

A double bond involves the sharing of two pairs of electrons, classically represented by the use of two bond lines between the atoms. A newer, more accurate understanding is that the two bonds are not identical, but the classical picture is adequate here. It also matches the design of most molecular model kits, in which two identical flexible plastic "bonds" are used to represent a double bond.

Rotation of the parts of a molecule that are connected by a double bond is not possible except under very high energy conditions. If the portions of the molecule at each of the two ends of the double bond are unsymmetrical, there will be two possible arrangements:

cis-2-butene   *trans*-2-butene

These are called **geometric isomers**. They are not structurally different (that is, they are not constitutional isomers). Geometric isomerism is one of two types of **stereoisomerism**. The other type, optical isomerism, will be studied later.

Note that when only three atoms are connected to an atom [like the carbon atoms in an alkene double bond (where the second carbon is one of the atoms attached)], the connected atoms again spread as far apart as possible, giving rise to 120° bond angles. Also, all six atoms that are a part of a double bond or are connected directly to it are coplanar. This contrasts with the four atoms directly connected with a tetrahedral atom, like the carbon atom in methane, where the bond angles are 109.5° and all atoms cannot be coplanar.

With alkenes having four or more carbons, we need a method of defining where the double bond is located. In the butane isomers above, the 2 in their names means that the double bond is located between the second and third carbons. These geometric isomers have a structural isomer 1-butene in which the double bond is between carbons 1 and 2.

Alkynes

The alkynes are hydrocarbons that contain one or more triple bonds between carbon atoms. This necessarily involves the loss of four hydrogen atoms per triple bond. Note here too there is a critical change in name suffix that distinguishes alkynes from alkanes. The alkyne triple bond is another type of functional group but is rarely of biological importance.

Cyclic Aliphatic Hydrocarbons

Alkanes and alkenes can be of the acyclic types shown above or they can be cyclic like cyclopentane, C_5H_{10}, and cyclohexene, C_6H_{10}:

The above compounds are shown in line-bond mode and in an alternative manner called **skeletal**. In skeletal mode, hydrogen atoms are not shown, just carbon and heteroatoms like oxygen, nitrogen, and chlorine. It is left to the reader to recognize that any of the four bonds per carbon atom that are not shown are used to hold a hydrogen atom. This method of drawing structures is advantageous because of the speed and clarity it provides for both writing and reading, especially for a large molecule like carotene mentioned above.

Formation of a ring requires removal of two hydrogen atoms and replacement of them with a new C-C bond. For example, the molecular formula C_5H_{10} is correct for either cyclopentane or a pentene (1-pentene or 2-pentene). Cyclopentane is saturated and so will not react with bromine but the pentenes will, being unsaturated.

It is possible to have a cycloalkyne, but to be stable its ring must contain nine or more carbon atoms because the array

$$C-C{\equiv}C-C$$

has to be linear, it will not bend. Its C-C-C bond angles are 180°, and therefore the rest of the ring must be long enough to span it.

The *cis/trans* system for indicating the relative positions of substituents on a double bond is rarely necessary in naming cycloalkenes because in the commonly occurring ring sizes (ones containing less than eight atoms), it is impossible to have a *trans* arrangement. However, we can use it to indicate the relative positions of two substituents on a cyclic compound:

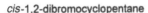

cis-1,2-dibromocyclopentane   *trans*-1,2-dibromocyclopentane

Note that for *each* substituent, a number is used to indicate its position. The lowest number(s) possible are chosen. If a double bond is present in a ring, it is always assigned to the 1 (to 2) position and then, now that the ring carbons have assigned numbers, substituents like a chlorine atom are given the lowest number possible.

[When it is unambiguous, the 1,2 location of the double bond is typically just "understood."]

3-chlorocyclopentene

AROMATIC HYDROCARBONS

Unique among the cyclic hydrocarbons is **benzene**, which with its derivatives constitutes the **aromatic** class of compounds. The term "aromatic" originated because the earliest examples were pleasant smelling natural products from plants like wintergreen, almond, and cinnamon. The term "aromatic" has since become an adjective used to describe compounds that display unexpected stability.

The central feature of these compounds is that they contain the benzene ring, found in its simplest, unsubstituted form in benzene itself, whose molecular formula is C_6H_6. Its structural formula was one of chemistry's great puzzles. The first one that proved useful is that of a six-membered ring having three double bonds separated by three single bonds. It was soon recognized that, if the carbons were numbered, there had to be an equilibrium involving interchange between these forms:

Linus Pauling, winner of two Nobel Prizes (one a Peace Prize for his efforts to ban nuclear weapons), found that the bonds between carbons in benzene are all of the same length, which is incompatible with the classical formula above because the double bonds in alkenes are shorter than single bonds. Pauling's suggested way of representing the uniform, intermediate length of the carbon-carbon bonds in benzene was to use the same familiar formulas shown above but to replace the pair of oppositely pointed arrows (which are still used to represent an equilibrium between two structures) with a new symbol, a **double-headed arrow.** This is used to indicate that the true structure is a single structure of intermediate nature, a hybrid structure.

Benzene's hybrid structure is now represented in these two alternative ways:

Pauling's method

Single picture method using line bond or skeletal

This single picture of the hybrid and Pauling's that uses two pictures of the two classical structures that each make a contribution to the hybrid are both used in chemical discussions because they each have advantages; the hybrid picture is quicker to draw but Pauling's system often gives more insight. The right-hand representation of the hybrid above is drawn in the skeletal, shorthand method; you must read in the hydrogens that are present at the ends of the undesignated bonds of carbon atoms, all of which are tetracovalent. If anything other than hydrogen is there, it must be specifically indicated, as here in toluene (methylbenzene) and 1,4-dichlorobenzene (which has been used, along with the bicyclic aromatic hydrocarbon naphthalene, also shown, to repel moths).

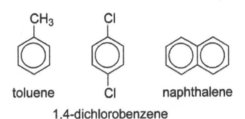

REACTIONS OF HYDROCARBONS

Oxidation

Natural gas is largely methane, and fuels like gasoline and Diesel oil consist of fairly complex mixtures of higher molecular weight liquid hydrocarbons, mostly alkanes but including also some alkenes and aromatics. Coal is a mixture of complex solid hydrocarbons. We all know that they burn and in the process produce much of the world's energy—and carbon dioxide. Their complete combustion is represented by the equation:

$$CH_4 + 2\,O_2 \longrightarrow CO_2 + 2\,H_2O + energy$$

But if there is inadequate oxygen for complete combustion, the oxidation proceeds only to poisonous carbon monoxide:

$$2\,CH_4 + 3\,O_2 \longrightarrow 2\,CO + 4\,H_2O + energy$$
$$\textit{TOXIC}$$

The above equations illustrate one of the two meanings of oxidation, the introduction of oxygen into a molecule, e.g., the further oxidation of carbon monoxide to carbon dioxide. In the experiment on ions (Experiment 3), we learned that oxidation is the removal of electrons, as in the oxidation of iodide ion (with its outer shell containing 8 electrons, completing its octet) to iodine (with only 7 electrons in its outer shell, forcing it to share one electron with one from a second iodine atom, forming molecular iodine, I_2). Actually, when a compound reacts with oxygen, it loses ownership in half of the bond electrons because oxygen's high electronegativity draws bond electrons towards itself. Consequently, both addition of oxygen and obvious removal of electrons are the same: the starting chemical loses (or loses full ownership) of electrons.

Hydrocarbons can undergo more limited oxidation than that in the combustion cases above. One of importance to many natural occurring materials, including foodstuffs and hydrocarbon fuels, is the slow but very damaging reaction of oxygen molecules with certain C-H bonds of alkenes, the ones adjacent to a double bond:

This reaction creates peroxides (compounds like HOOH, hydrogen peroxide) that explain rancidity in old samples of unsaturated fats and poor performance in old vehicle fuels.

Another is the introduction of two oxygen atoms into an alkene double bond by its reaction with permanganate ion, MnO_4^-. This is one of the common tests used to distinguish alkenes from alkanes and aromatic hydrocarbons, which do not react. Oxidation of the colorless alkene reduces the intensely purple, soluble permanganate ion to a precipitate of brown-black manganese dioxide:

$$3 H_2C{=}CH_2 + 2 MnO_4^- + 4 H_2O \longrightarrow 3 H_2C{-}CH_2 + 2 MnO_2\downarrow + 2 HO^-$$

$$\underset{\substack{\text{purple}\\\text{aq.soln.}}}{} \qquad\qquad \underset{\substack{\text{OH OH}}}{} \underset{\substack{\text{brown-black}\\\text{precipitate}}}{}$$

Another chemical reaction of alkenes is with bromine; it is another test useful for distinguishing alkenes from alkanes and aromatic hydrocarbons:

$$H_2C{=}CH_2 + \underset{\substack{\text{red-}\\\text{brown}}}{Br_2} \longrightarrow \underset{\substack{\text{Br Br}\\\text{colorless}}}{H_2C{-}CH_2}$$

The failure of benzene and related aromatic compounds to react with either permanganate ion (usually employed as the potassium salt, $KMnO_4$) or bromine, Br_2, was very puzzling in view of the earlier belief that they had three double bonds in their rings. This lower reactivity exemplifies the unusual stability that is now recognized as the defining characteristic of aromatic compounds.

Using more powerful methods than used in the above tests for alkene double bonds, even alkanes and aromatic hydrocarbons show reactivity. For example, bromine will react with alkanes at high temperature or in ultraviolet (high energy) light and also with aromatic hydrocarbons if an acidic catalyst is employed.

MOLECULAR MODELS

The "ball-and-stick" molecular model set you will use in this experiment consists of colored wooden or plastic spheres, representing atoms, drilled to receive connecting bonds. Your instructor will furnish the color code for your model set. Carbon atom spheres have four drilled holes that represent the four covalent bonds that carbon atoms always form. Hydrogen atoms form one bond, bromine atoms form one bond, and oxygen atoms form two bonds; spheres for these atoms usually have the appropriate number of holes.

Single covalent carbon-to-carbon bonds are represented by wooden or plastic pegs that can be inserted into the holes of the atoms. Some model sets use shorter pegs to represent bonds between carbon and any other element. Double bonds are created either with two bent-pegs or with two bendable spring-pegs.

The structures you will assemble in this experiment closely approximate the geometry of the molecules they represent. Two structures are identical if they are superimposable—that is, if one structure can be placed "on top" of another so that all the colored spheres coincide.

Procedure

I. MOLECULAR MODELS: ISOMERISM

A. Methane

1. Construct two models of the methane (CH_4) molecule. Superimpose one model on top of the other to show they are identical structures. Sketch the three-dimensional structure of the methane molecule. Note the 109.5° tetrahedral bond angles of the three-dimensional methane molecule.

2. Remove one of the hydrogen atoms from one of the structures and replace it with a bromine atom. Place the structure aside for reference.

3. On the second molecule of methane you constructed in Step 1, replace one hydrogen atom with a bromine atom. Compare this structure, by attempting to superimpose one structure on top of the other, with the reference model that you constructed in Step 2.

4. Place the reference model aside. On the second model, interchange the bromine with any other hydrogen. Compare this new structure with the reference model by attempting to superimpose the models.

B. Ethane

1. Construct two models of the ethane (C_2H_6) molecule. Sketch the three-dimensional structure of the model.

2. On each of the two ethane models, replace one of the hydrogen atoms with a bromine atom. Compare their structures by attempting to superimpose the models, and place one of the C_2H_5Br models aside for reference.

3. On the second C_2H_5Br structure, interchange any one of the five hydrogen atoms with the bromine atom. Compare the structure to the reference molecule by attempting to superimpose the models.

4. Repeat Step 3 several times; after each interchange, compare the structure to the reference molecule.

5. On one of the C_2H_5Br models, rotate the carbon atoms around the bond joining them so that the atoms assume a "staggered" conformation as shown in Figure 5-1. Draw this conformation using a Newman projection.

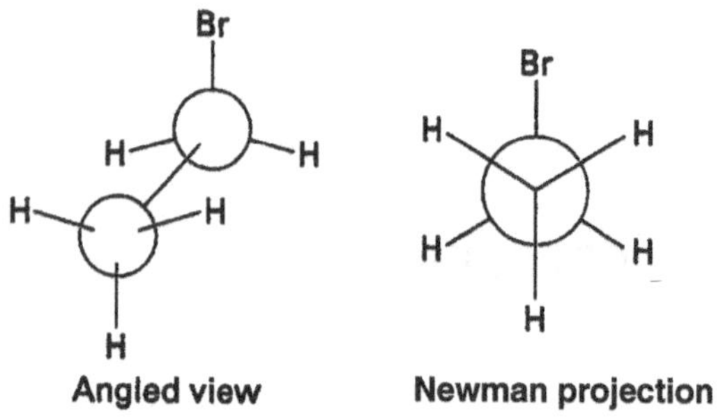

Figure 5-1 Model of bromoethane in "staggered" position.

6. Hold the carbon atom attached to the three hydrogens in a fixed position and rotate the second carbon atom (attached to the bromine) one-sixth of a turn until an eclipsed conformation is obtained. Draw its Newman projection.

7. Assemble models of as many different isomers of $C_2H_4Br_2$ as you can, comparing their structures by attempting to superimpose the models. Write the structural formula and the IUPAC name of each isomer.

C. Propane

1. Construct a model of a propane (C_3H_8) molecule.

2. Draw the line-bond structural formula of propane.

3. Construct models of as many different isomers of C_3H_7Br as possible. Compare the structures by attempting to superimpose the models. Write the structural formula of each. Write the IUPAC name of each isomer.

D. Butane and Pentane

1. Construct models of as many different isomers of C_4H_{10} as you can. Write the skeletal structural formula and the IUPAC name of each.

2. Construct models of as many different isomers of C_5H_{12} as possible. As for butane, write the structural formula and name of each isomer.

E. Ethylene (Ethene)

1. Construct a model of ethylene (C_2H_4). Use two bent-pegs or two bendable spring-pegs (depending on the type of model kit you are using) between the carbon atoms to represent the double bond. Write the structural formula of ethylene. Try to rotate the two carbon atoms as you did in Step B5.

2. Construct models of as many different isomers of $C_2H_2Br_2$ as possible. Write their structural formulas and names.

F. Cycloalkanes

1. Construct models of as many different five-membered ring structural (constitutional) isomers of $C_5H_8Br_2$ as possible. Write the structural formula and name of each. Studying these models, decide how many of these structural isomers can exist in geometrically isomeric forms.

2. Construct two models of cyclohexane (C_6H_{12}). Attempt to rotate any two adjacent carbon atoms. Since they are in a ring, rotation is limited but even the slight rotation possible is very significant in explaining relative stability and reactivity. It permits a six-membered ring to assume two principal conformations. Bend opposite carbons (call them carbons 1 and 4) up for one conformation, and (using the other model) one up and one down for the second. Draw a side view of each conformation (in these, have carbons 2 and 3 overlapping carbons 5 and 6).

   Thinking in terms of Newman projections, sight down the C.2-C.3 and C.5-C.6 bonds in each of the models and predict which conformation will be more stable (staggered conformations are always more stable than eclipsed ones).

   One conformation is called the **chair form** (it looks more like a chaise lounge than a simple chair) and the other the **boat form**. Predict which is which and assign these names to your drawings.

II. PROPERTIES OF HYDROCARBONS

A. Solubility and Density

1. Place 8 drops of the following hydrocarbons into six numbered test tubes as follows:
   No. 1 and 4: heptane
   No. 2 and 5: 1-hexene
   No. 3 and 6: toluene

2. Add 2 mL of deionized water to test tubes No. 1–3; cork and shake vigorously. Classify the compounds as insoluble (I) or soluble (S) in water.

3. Observe and record whether each hydrocarbon is more or less dense than water.

4. Add 2 mL of ligroin (a low-boiling fraction of crude oil) to test tubes No. 4–6; cork (must be dry) and shake. Classify the compounds as I or S.

5. **Disposal:** Pour the liquids into the "Waste Hydrocarbon" container.

> **CAUTION: Heptane, 1-hexene, toluene, and ligroin are flammable. As a general rule, consider all organic compounds to be combustible. The vapors of many organic compounds are believed to be harmful to your health; physicians may advise pregnant women to avoid breathing vapors of organic compounds. To reduce the concentration of vapors in the room, the hydrocarbons used in this experiment should be handled only in a fume hood.**

B. Tests for Unsaturation

In both of these tests, if reaction occurs, it represents a limited, partial oxidation (combustion contrasts in being a complete oxidation).

B.1 Permanganate Test

1. Put 10 drops of heptane into a clean, dry test tube. Note its color.

2. Note the intense purple color of 1% aqueous potassium permanganate solution, the test reagent to be used.

3. Add 20 drops of acetone; then add 1% aqueous potassium permanganate solution dropwise until the purple color persists. Stir with a glass rod after each addition. Count and record the number of drops required. (Note that if the purple color disappears or diminishes, a brown-black precipitate appears. It is the product of reduction of the permanganate ion to manganese dioxide.)

4. Repeat Steps 1-3 with 1-hexene and with toluene.

5. **Disposal:** Pour the contents of the three test tubes into the "Waste Manganese Dioxide" container. Clean the test tubes as directed by your instructor.

B.2 Bromine Test

> **The bromine solution may burn and blister the skin. Bromine vapors are irritating to eyes and lungs. Bromine solutions should be handled only in a fume hood. Wear disposable laboratory gloves to handle this solution.**

1. **Perform this test in a fume hood.**

2. Put 10 drops of heptane into a clean, dry test tube. Note its color.

3. Note the red-brown color of bromine solution (bromine dissolved in dichloromethane). Add the bromine solution drop by drop into the test tube until a faint color of bromine persists. **CAUTION: Gently stir the test tube after the addition of each drop.** Count the number of drops required to produce a faint bromine color. **CAUTION: Do not add more than 20 drops.**

4. Repeat Steps 1–3 with 1-hexene and with toluene.

5. **Disposal:** Pour the 3 hydrocarbon solutions from the test tubes into the "Halogenated Solvent Waste" container. Clean the test tubes as directed by your instructor.

Prelab Questions for Experiment 5

_____ 1. Two structures are identical when
   a. they have the same molecular formula
   b. they have the same combustion products
   c. they can be superimposed on each other

_____ 2. Toluene is classified as
   a. an alkane
   b. an alkene
   c. an alkyne
   d. an aromatic hydrocarbon

_____ 3. Cyclopentane is classified as
   a. an unsaturated compound
   b. a saturated compound
   c. an alkene
   d. an alkyne

_____ 4. Several drops of a bromine solution were added to a liquid. The red-brown color of the bromine immediately disappeared. The compound may have been
   a. a saturated hydrocarbon
   b. a cycloalkane
   c. an aromatic hydrocarbon
   d. an alkane
   e. an alkene

5. a. Write the skeletal structural formula of heptane.

   b. What is the molecular formula of heptane?

6. a. Write the line-bond structural formula of 1-hexene.

   b. What is the molecular formula of 1-hexene?

7. a. Write the structural formula of toluene using the Pauling method.

   b. What is the molecular formula of toluene?

8. What is the molecular formula of:

   a. An alkane that has three carbon atoms?

   b. An alkane that has n carbon atoms?

   _____

   c. A cyclic alkene that has five carbon atoms and one double bond?

   _____

9. Complete and balance the following equations:

   a.

   $+ Br_2 \longrightarrow$

   b. $C_5H_{12} + 8 O_2 \rightarrow$

Name ______________________________ Section ____________ Date ______________

Report for Experiment 5

I. MOLECULAR MODELS: ISOMERISM

A. Methane

1. Sketch of the methane model:

2. Draw the structural formula of methane using the 3-D method that employs wedge bonds:

3. Are all the hydrogens in methane **equivalent**; that is, do you obtain the same structure by replacing any one of the hydrogens with bromine?

   ____

B. Ethane

1. Sketch of the ethane model:

2. Draw the structural formula of ethane using wedge bonds:

3. Are all the hydrogens in ethane equivalent; that is, do you obtain the same molecule by replacing any hydrogen with bromine?

   ____

4. Draw Newman projections of C_2H_5Br conformations:

a. The one studied in Step 5 (staggered)

b. The one studied in Step 6 (after one-sixth of a turn; eclipsed)

c. Do the above drawings of C_2H_5Br represent the same or different structures?

d. Do the following structural formulas represent the same or different molecules?

$$
\begin{array}{ccc}
\text{H} & \text{Br} \\
| & | \\
\text{H}-\text{C}-\text{C}-\text{H} \\
| & | \\
\text{H} & \text{H}
\end{array}
\qquad
\begin{array}{ccc}
\text{H} & \text{H} \\
| & | \\
\text{H}-\text{C}-\text{C}-\text{Br} \\
| & | \\
\text{H} & \text{H}
\end{array}
\qquad
\begin{array}{ccc}
\text{H} & \text{H} \\
| & | \\
\text{H}-\text{C}-\text{C}-\text{H} \\
| & | \\
\text{H} & \text{Br}
\end{array}
$$

5. a. Structural formulas and names of isomers of $C_2H_4Br_2$:

b. Draw the Newman projection of the most stable conformation of 1,2-dibromoethane (the one that has the largest atoms as far apart as possible).

Name ______________________ Section ____________ Date ____________

C. Propane

1. Line-bond formula of propane:

2. Line-bond formula and name of each isomer of C_3H_7Br:

D. Butane and Pentane

1. Skeletal formulas and names of isomers of C_4H_{10}:

2. Skeletal formulas and names of isomers of C_5H_{12}:

E. Ethylene (Ethene)

    1. a.   Line-bond formula of ethylene:

        b.   Do the carbon atoms have freedom of rotation?

    2. Structural formulas and names of isomers of $C_2H_2Br_2$:

F. Cycloalkanes

    1. a.   Skeletal formulas and names of all possible structural isomers of $C_5H_8Br_2$ that contain a five-membered ring:

        b.   Circle the formulas of any of these isomers that can exist in geometrically isomeric forms.

    2. a.   Draw the side views of the two principal conformers of cyclohexane.

        b.   Circle the side view of the conformer you predict will be more stable.

        c.   Write your predicted name (chair vs. boat) beside their side view drawings.

Name ______________________ Section ____________ Date ____________

II. PROPERTIES OF HYDROCARBONS

A. Physical (Solubility and Density)

1. Data:

Hydrocarbon	*Solubility (I or S) in*		*Density Compared to Water (M or L)*
	Water	*Ligroin*	
Heptane			
1-Hexene			
Toluene			

2. In what type of solvent, polar or nonpolar, are hydrocarbons soluble?

____

B. Tests for Unsaturation

1. Permanganate Test Data:

Hydrocarbon	*Number of Drops Decolorized*	*Brown-black Precipitate (Y or N)*	*Unsaturated (Y or N)*
Heptane			
1-Hexene			
Toluene			

2. Which class of hydrocarbon (alkane, alkene, or aromatic) was most reactive?

____

3. Write a skeletal formula for the product of reaction of $KMnO_4$ with l-hexene.

4. Bromine Test Data:

Hydrocarbon	Number of Br_2 Solution Drops Decolorized
Heptane	
1-Hexene	
Toluene	

5. Which class of hydrocarbon was most reactive with bromine?

_____

6. Write a balanced equation for each hydrocarbon that reacted with bromine. Use line-bond structural formulas (not molecular formulas) in the equation(s).

Related Questions for Experiment 5

1. Draw structural formulas for all the isomers of C_6H_{14}. Name each isomer.

2. Draw structural formulas for all the isomers of C_4H_6. You need not name them. [There are seven isomers (two have triple bonds).]

Experiment 6
Properties of Water

Reference: *General, Organic, and Biological Chemistry: An Integrated Approach, 4th ed., Chapters 4 and 6.*

Purpose: You will investigate several properties of water including its viscosity, freezing point, boiling point, and surface tension. The solvent properties of water will be examined and you will study saturated, unsaturated, and supersaturated solutions.

Introduction

THE SOLVENT PROPERTIES OF WATER

Water is the most important biological solvent. Blood serum, urine, perspiration, and lymph are examples of water solutions in the body. The highly polar nature of water allows a large number of solutes to dissolve in it. The rule "like dissolves like" means that most ionic compounds, as well as polar molecular compounds, will dissolve in water. Nonpolar molecular substances, such as gasoline, do not dissolve in water and do not form a solution.

If a substance dissolves in water to form a solution, the mixture will be **homogenous**—it will appear totally uniform with no visible boundaries between one substance and another. If a substance does not dissolve in water, the mixture will be **heterogeneous**—there will be a boundary between the two phases. A cloudy mixture contains tiny pieces of undissolved, suspended solid and is not a solution.

SATURATED, UNSATURATED, AND SUPERSATURATED SOLUTIONS

A **saturated solution** is one in which the solvent has dissolved all the solute it can at a particular temperature. In a saturated solution, dissolved solute and undissolved solute are in equilibrium. The rate at which solid solute dissolves equals the rate at which dissolved solute crystallizes.

If you add many tablespoons of sugar (with stirring) to a cup of water, eventually there will be solid, undissolved sugar on the bottom of the cup. The dissolved sugar is in equilibrium with the undissolved sugar; the solution is saturated with sugar.

$$\text{sugar}_{\text{(undissolved)}} \rightleftharpoons \text{sugar}_{\text{(dissolved)}}$$

If there is only a small amount of solid, undissolved sugar, heating the solution will dissolve the remaining solid. You could stir in a little more sugar and it would dissolve in the hot, unsaturated solution. When sufficient solid sugar has been added, a new saturated solution will be formed. Most liquid and solid solutes become more soluble with increased temperature. A solution may be saturated at one temperature and unsaturated at an elevated temperature.

An **unsaturated solution** contains less solute than the solvent can dissolve at a given temperature. If a little more solute were added, it would dissolve. In an unsaturated solution, dissolved solute and undissolved solute are not in equilibrium—the rate of dissolving is greater than the rate of crystallization.

Cooling of a hot saturated solution usually results in release of enough solute to reduce the liquid phase concentration to that of a saturated solution at the new, lower temperature. If the solute is a solid at the lower temperature, it usually will appear in crystalline form. When this procedure (cooling of a hot saturated solution) yields crystals, it is called **recrystallization**. The recrystallized solute is more pure than it was originally because the cooling rarely makes the solution reach the solubility limit of the small amounts of the impurities that may be present so they stay in solution. The purified crystals can be separated by filtration. The filtrate contains the impurities but also remains saturated with the main solute, a loss that is the price of purifying the recrystallized portion.

A **supersaturated solution** is an unstable solution in which the solvent has dissolved more solute than it theoretically can at that temperature. The molecules of a liquid solution are disordered and unorganized; the molecules of a solid crystal are highly organized. To crystallize, a supersaturated solution needs a point of molecular orderliness around which the disordered molecules can crystallize. A disturbance, such as sudden shaking or addition of a tiny piece of solid solute, can furnish the molecular order and cause some of the undissolved solute to form solid crystals. When it crystallizes in this way, the unstable supersaturated solution loses its excess solute and becomes a saturated solution. Honey is an example of a supersaturated sugar solution. Honey sometimes crystallizes as the supersaturated solution turns to a more stable saturated solution. To reform the supersaturated solution, you must heat the solid honey until all solid dissolves, then cool slowly. Most substances crystallize when a saturated solution is cooled, but a few, like honey, cool without crystallizing and become supersaturated.

Supersaturated solutions cannot be formed by stirring additional solute into a saturated solution—the added solute will not dissolve. A supersaturated solution can be made by carefully cooling a hot, saturated solution.

One method to determine whether a solution is saturated, unsaturated, or supersaturated is to add a tiny piece of solid solute. One of three fates awaits the seed crystal—it can dissolve in an unsaturated solution and get smaller, it can stay the same size because a saturated solution cannot dissolve any more solute, or it can get larger because it triggered the crystallization of a supersaturated solution.

Original Solution	Added Solute Crystal	Resultant Solution
Unsaturated	Gets smaller (dissolves)	Unsaturated or saturated depending on how much solute was added
Saturated	Stays the same size	Saturated
Supersaturated	Gets larger	Saturated

In this experiment you will investigate the solutions of two solutes, sodium chloride and sodium thiosulfate. Sodium chloride is, of course, common table salt. Sodium thiosulfate ($Na_2S_2O_3$) is also known as "hypo" and is used as a fixer in photography. Sodium thiosulfate removes chlorine from solutions in the bleaching of paper pulp. In medicine, it is an antidote for cyanide poisoning.

SURFACE TENSION

Have you ever seen "water bugs" walking across the surface of water in a pond? Water has a kind of "skin" on its surface that holds up the bug. The water molecules on the surface resist being pushed apart from each other by the bug's legs—this resistance to expansion of the surface of a liquid is called **surface tension**. Surface tension is caused by unbalanced forces of attraction. Through hydrogen bonding, water molecules at the surface are pulled inward toward interior water molecules. The force is unbalanced because there are no attractive forces pulling outward on the surface molecules. As a result, any liquid surface becomes as small as possible—this causes the spherical shape of water drops and soap bubbles.

All liquids have surface tension. Water has an unusually high surface tension because its hydrogen bonding is stronger than other types of intermolecular attraction. Detergents weaken the surface tension of water when the detergent molecules move between the water molecules and weaken their attraction for one another.

VISCOSITY

Viscosity is a measure of how easily a liquid flows. High-viscosity liquids like honey resist flow; low-viscosity liquids flow easily. The viscosity of a liquid is affected by changes in temperature—you will investigate the direction of that change in this experiment. Motor oils are available in different viscosities. The viscosity of multi-viscosity motor oils, such as 10W-30, is less affected by temperature changes than single-viscosity oils (such as straight 30 grade viscosity).

You will study the viscosity of several common liquids. Isopropyl alcohol ($CH_3CHOHCH_3$; a 70% solution in water is known as rubbing alcohol) is medically useful as an antiseptic. Ethylene glycol ($HOCH_2CH_2OH$) is the main ingredient in automotive antifreeze. Glycerol ($HOCH_2CHOHCH_2OH$), also called glycerin, is a moisturizer in many foods and cosmetics. Glycerol is an important part of triglycerides (fats), its triesters derived from fatty acids. Nitroglycerin, the explosive in dynamite and medically useful as a vasodilator, is made from glycerin. Note the increasing numbers of hydroxyl groups (O-H) in these similarly sized molecules. That permits increasing opportunities for hydrogen bonding between molecules, which increases their hold on each other. This is apparent from their relative viscosity, which you will study, and from their relative boiling points, which are 82, 198, and 290 °C, respectively.

Dashed lines represent **hydrogen bonds**. They occur when a hydrogen atom on one highly electronegative atom is attracted to another.

Hydrogen bonds are the strongest of the intermolecular attractions but are only about a tenth as strong as the average covalent bond.

FREEZING POINT

The **freezing point** of a substance is the temperature at which the solid and liquid are in equilibrium. The freezing point of a liquid is the same temperature as the **melting point** of the solid. The **ice point** is the temperature of an ice-water mixture and is approximately the same as the freezing point, but is more easily measured.

You will measure the ice point of both ice-water and of a salt-water solution prepared by adding rock salt to ice-water. Rock salt is naturally occurring, unpurified sodium chloride. Rock salt is used on streets to melt ice. An even better use of rock salt is on ice in making home-made ice cream. In making ice cream we want the ice to absorb heat from the cream so that the cream will freeze. Making ice melt and absorbing heat are different viewpoints of the same phenomenon. The melting of ice is an equilibrium process—ice is melting and water is freezing at the same rate. When ice melts, heat is absorbed from the surroundings; when water freezes, heat is released to the surroundings.

$$\text{ice (at } 0°C) \ + \ \text{heat} \ \rightleftharpoons \ \text{liquid water (at } 0°C)$$

The **heat of fusion** is the amount of heat required to change one gram of ice (or another solid) at its melting point from a solid to a liquid. Compared to other substances, water has a very high heat of fusion.

When salt, or any other solute, is added to water, the freezing point of the solution is lowered—this is called **freezing point depression**. With added salt, the ice is melting faster than water is freezing; the faster melting ice absorbs heat from its surroundings. Ice on the streets is melted because the salt water solution no longer freezes at 0°C but at a lower temperature—the exact temperature depends on the concentration of the salt solution. A mixture of 33 grams of salt and 100 grams of water will stay liquid until the temperature drops to −21°C. Calcium chloride, $CaCl_2$, lowers the freezing point of water more than sodium chloride, NaCl, because it is more soluble in water (and because each of the three ions, Ca^{2+}, Cl^-, and Cl^-, in the formula unit contributes to the freezing point depression).

It is the freezing point lowering caused by addition of ethylene glycol to car radiator fluid that makes it useful as "antifreeze." Keeping radiator water from freezing is important because of another property of water: it expands on freezing, which can crack engine blocks (or any other container). Also, it is this expansion on freezing that lowers the density of ice enough that it floats on water.

BOILING POINT

The **boiling point** of a liquid is the temperature at which the vapor pressure of the liquid equals the atmospheric pressure. Since atmospheric pressure varies from location to location and from day to day, the boiling point will change as the pressure varies. The **normal boiling point** of a liquid is the temperature at which the vapor pressure of the liquid equals 1 atmosphere. Water has a normal boiling point of 100°C. At atmospheric pressures less than 1 atmosphere, water will boil at a lower temperature; likewise, at a higher atmospheric pressure it will boil at a temperature greater than 100°C. Pressure cookers cook rapidly because the boiling water is hotter than 100°C; the vapor pressure of the water in the pot has to reach a higher value to become equal to the atmospheric pressure inside the pressure cooker.

Water has an amazingly high boiling point for a compound that has such a low molecular weight, 18 amu. Molecular weight (MW) is the universally important determinant of boiling point: boiling point generally increases as MW increases. Compare the boiling points of methane (MW 16, bp −182°C) and ethane (MW 30, bp −89°C). In the alkane series, you have to go to heptane (MW 100, bp 98°C) before the boiling point becomes about equal to that of water. The explanation lies in the extreme opportunity for hydrogen bonding in water, which is essentially all hydroxyl group.

At temperatures below the boiling point, evaporation occurs only from the surface of the liquid. During boiling, vapor forms throughout the liquid; as the vapor bubbles rise through the liquid, they create the turbulence associated with boiling. Heat must be constantly added to maintain boiling. The heat (**heat of vaporization**) changes the physical state of the water but not its temperature.

$$\text{liquid water (at 100°C)} \quad + \quad \text{heat} \rightleftarrows \text{water vapor (at 100°C)}$$

The boiling point of a solution is higher than the boiling point of the solvent—that is, adding solute to a solvent raises the solvent's boiling point. This is referred to as **boiling point elevation**. This phenomenon is another reason that addition of ethylene glycol to radiator fluid is advantageous. It permits operation at higher engine temperature, something that increases fuel efficiency.

An important thing to know about liquids that are being heated is that they can sometimes go to an unstable, superheated state, one in which they reach temperatures above their boiling point. Then, just as with supersaturated solutions, they can suddenly change, which here results in "bumping." It is an abrupt conversion of excess heat by what can be an explosive release of hot vapor that can also splash hot liquid. To avoid this, it is common practice to add a "boiling chip," a piece of porous stone that contains air in its pores that on warming in the liquid expands, forming bubbles that catalyze smooth conversion of the liquid to vapor at its boiling point. **It is essential** that a boiling chip be added **well below** the boiling point; if the liquid has already gotten superheated, addition of the chip will guarantee instantaneous, violent bumping!

Procedure

I. SATURATED, UNSATURATED, AND SUPERSATURATED SOLUTIONS

Start Part B first to allow for the one-hour cooling required.

A. Saturated Sodium Chloride Solution

1. Place 3 mL of a saturated sodium chloride solution into each of two 10 × 75-mm test tubes (the smallest of the 3 sizes). Label them A and B.

2. Add a few sodium chloride crystals to test tube A, stopper, and shake vigorously. Note if the crystals dissolve.

3. Add a few sugar crystals to test tube B, stopper, and shake vigorously. Observe.

B. Solutions of Sodium Thiosulfate

1. Into a 25 × 150-mm test tube (the largest of the 3 sizes) place 5 g of sodium thiosulfate, $Na_2S_2O_3$. Add 15 drops of water. Note if the solid dissolves.

2. Add 200 mL of water to a 400-mL beaker and prepare a boiling-water bath as shown in Figure 0-20 and 0-21 on page 20. Place the test tube in the hot-water bath. The test tube can be added before the water reaches a boil. Heat the test tube until a clear solution is obtained; all crystals must be dissolved.

3. Add a single, small crystal of sodium thiosulfate to the hot solution. Observe the crystal as it settles to the bottom of the test tube. Wait 1 minute before proceeding with Step 4.

4. Carefully remove the test tube from the bath and allow it to cool for one hour. Do not agitate the test tube. Save the hot-water bath for Parts II and III.

5. After one hour of cooling, add a single, small crystal of sodium thiosulfate to the solution. Observe the crystal as it settles to the bottom of the test tube. Feel the outside of the test tube for temperature changes.

6. Although it may be hard to see, there is a solution in the test tube. Add a single crystal of sodium thiosulfate to the test tube. Observe what happens to the crystal.

7. **Disposal:** Heat the test tube in boiling water to dissolve the solid. Pour the hot solution into the "Waste Inorganics" container.

II. VISCOSITY

1. Heat the hot-water bath from Part IB to boiling.

2. Using another 400-mL beaker, prepare an ice-water bath containing 200 mL of crushed ice and 20-mL of water.

3. Place eight 16 × 150-mm test tubes in a rack, label them as indicated below, and add the following reagents. Observe any differences in their viscosities as you fill the test tubes. **CAUTION: isopropyl alcohol, ethylene glycol, and glycerol are flammable.**

   No. 1A: 5 mL of isopropyl alcohol (2-propanol)
   No. 1B: 5 mL of isopropyl alcohol
   No. 2A: 5 mL of ethylene glycol (1,2-ethanediol)
   No. 2B: 5 mL of ethylene glycol
   No. 3A: 5 mL of glycerol (1,2,3-propanetriol)
   No. 3B: 5 mL of glycerol
   No. 4A: 5 mL of water
   No. 4B: 5 mL of water

4. Place the A test tubes in the ice-water and the B test tubes in the boiling-water bath.

5. After approximately three or four minutes remove each No. 1 test tube from its bath and shake gently. Try to pour each liquid. Note any differences.

6. Repeat Step 5 with No. 2, No. 3, and No. 4 test tubes. Then put contents of the 1-3 tubes in the "Waste Organics" container.

7. Save the boiling-water bath for Part III.

III. THE BOILING POINT OF WATER

1. Heat the hot-water bath you used in Part II until the water boils. Make sure the water has come to a full, rolling boil.

2. Insert a thermometer into the boiling water, and when the thermometer displays a constant temperature, record the temperature to the precision allowed by the thermometer.

IV. THE ICE POINT OF WATER AND OF AN AQUEOUS SOLUTION

1. Use a 250-mL beaker to prepare a new ice bath containing 150 mL of crushed ice and 10 mL of water.

2. Immerse a thermometer in the water-ice slurry and stir **with a stirring rod** for at least two minutes. Leave the thermometer in the ice bath but be careful not to jar it with the stirring rod. When the column in the thermometer stops moving or the reading on a digital thermometer is constant, read the thermometer to the nearest 0.1°C.

3. Continue stirring and repeat the thermometer measurement twice more at two minute intervals.

4. Add 20 grams of rock salt (NaCl), thoroughly stir for two minutes, and take a temperature reading.

5. Continue the thorough stirring and temperature measurement twice more at two minute intervals. Note if the salt melts the ice.

V. SURFACE TENSION

1. Fill a shallow, soap-free metal pan with water to a depth of about 1 or 2 cm.

2. Float a small piece of tissue paper (about 3 cm × 3 cm) on the surface of the water. Rest a soap-free straight pin on the tissue paper.

3. Slowly squirt liquid dish soap into the water on the opposite side of the pan from the pin. Observe the behavior of the pin.

VI. SOLUBILITIES OF COMMON SUBSTANCES IN WATER

1. Place seven 16 × 150-mm test tubes in a rack, number them, and add the following substances ("A small amount" means adding, from the end of a microspatula, an amount approximately equal in volume to a small pea):

   No. 1: A small amount of potassium nitrate (KNO_3)
   No. 2: A small amount of table sugar (sucrose, $C_{12}H_{22}O_{11}$)
   No. 3: 20 drops of vegetable oil (a mixture, typically $C_{39}H_{66}O_6$)
   No. 4: A small amount of barium sulfate ($BaSO_4$, an X-ray medium)
   No. 5: 20 drops of isopropyl alcohol (C_3H_7OH; **CAUTION**, flammable; rubbing alcohol is a 70% aqueous solution).
   No. 6: 20 drops of 1-pentanol ($C_5H_{11}OH$; **CAUTION**, flammable)
   No. 7: 20 drops of glacial acetic acid (CH_3COOH; **CAUTION**, strong odor and corrosive; vinegar is a 5% aqueous solution).

2. Add 5 mL of deionized water to each of the test tubes, stopper them, and mix for a minute or two. If the mixture is homogeneous (if there is only one phase), the substance is soluble in water. If there is more than one phase, the substance is insoluble in water. Use S or I on the Report Sheet to indicate solubility or insolubility.

3. **Disposal:** Place the contents of tubes 1 and 4 in the "Waste Inorganics" container and the contents of the others in the "Waste Organics" container. Clean the test tubes as directed by your instructor.

Properties of Water

Prelab Questions for Experiment 6

1. When a sodium chloride crystal is dropped into a solution, it gets smaller in size and disappears. Was the solution saturated, unsaturated, or supersaturated with sodium chloride?

   _____

2. When a saturated solution is heated, does it usually remain saturated, become unsaturated, or supersaturated?

   _____

3. When two liquids are mixed together in a test tube, only one phase results. Are the liquids soluble or insoluble in each other?

   _____

4. The following substances are used in Part VI of this experiment. Use I or M to classify each as an ionic compound or a molecular compound. Hint: see Appendix.

   _____ potassium nitrate, KNO_3

   _____ sugar, $C_{12}H_{22}O_{11}$

   _____ cooking oil, $C_{39}H_{66}O_6$

   _____ barium sulfate, $BaSO_4$

   _____ isopropyl alcohol (2-propanol), $CH_3CHOHCH_3$

   _____ 1-pentanol, $CH_3(CH_2)_4OH$

   _____ acetic acid, CH_3COOH

5. Are ionic compounds usually soluble or insoluble in water?

   _____

6. If a molecular compound dissolves in water, is it a polar molecular compound or a nonpolar molecular compound?

   _____

Report for Experiment 6

I. SATURATED, UNSATURATED, AND SUPERSATURATED SOLUTIONS

A. Saturated Sodium Chloride Solution

    1. a. Did the sodium chloride crystals dissolve when you added them to test tube A?

_____yes_____

       b. Explain your observations.

_____As I was shaving the solution the_____
_____NaCl dissolved_____

    2. a. Did the sugar crystals dissolve as you added them to test tube B? _____

       b. Explain your observations.

_____

_____

B. Solutions of Sodium Thiosulfate

    1. a. What happened in Step 3 when you added a crystal of sodium thiosulfate to the hot sodium thiosulfate solution?

_____

       b. Before you added the crystal in Step 3, was the hot sodium thiosulfate solution saturated, unsaturated, or supersaturated?

_____

       c. After you added the crystal in Step 3, was the hot sodium thiosulfate solution saturated, unsaturated, or supersaturated?

_____

    2. a. What happened in Step 5 when you added a crystal of sodium thiosulfate to the cold sodium thiosulfate solution?

_____

b. Before you added the crystal in Step 5, was the cold sodium thiosulfate solution saturated, unsaturated, or supersaturated?

_____

c. After you added the crystal in Step 5, was the cold sodium thiosulfate solution saturated, unsaturated, or supersaturated?

_____

d. Is the crystallization of sodium thiosulfate an exothermic or an endothermic process?

_____

3. a. What happened in Step 6 when you added a crystal of sodium thiosulfate to the hard-to-see sodium thiosulfate solution?

_____

b. Before you added the crystal in Step 6, was the hard-to-see sodium thiosulfate solution saturated, unsaturated, or supersaturated?

_____

c. After you added the crystal in Step 6, was the hard-to-see sodium thiosulfate solution saturated, unsaturated, or supersaturated?

_____

II. VISCOSITY

1. Which of the four reagents (isopropyl alcohol, ethylene glycol, glycerol, or water) is most viscous at room temperature?

_____

2. How was the viscosity of the four reagents affected by lowering the temperature?

_____

3. How was the viscosity of the four reagents affected by increasing temperature?

_____

III. THE BOILING POINT OF WATER

1. Data. Observed boiling point of water: _____

2. Explain why your measured boiling point might be different from the accepted boiling point of water.

   _____

   _____

IV. THE ICE POINT OF WATER AND OF AN AQUEOUS SOLUTION

1. a. Data. Ice Point of Water:

	Trial 1	Trial 2	Trial 3	Average
Ice point of water				

   b. In this procedure, you measured the temperature of the ice-water mixture every two minutes as it sat at room temperature. If you continued stirring the ice-water mixture, when would the temperature of the ice-water increase?

   _____

2. a. Data. Ice Point of the Salt Solution:

	Trial 1	Trial 2	Trial 3	Average
Ice point of salt solution				

   b. What is the approximate concentration of your salt solution in units of weight percent? _____ (In calculating this, estimate the weight of the ice and water that you used.)

   c. Look up and record here the reported ice (freezing) point for a solution of your concentration. (In a copy of the *Handbook of Chemistry and Physics*, look in the index for "freezing point depression." This will lead you to a section in which data tables for many different solutes are listed in alphabetical order. The first column in each is headed "A% by wt." Another, midway to the right, is headed "Δ°C" and gives the reported freezing point depression, which for water is the ice point.) _____

V. SURFACE TENSION

1. Why did the straight pin float on the surface of the water?

   _____

2. What happened to the pin as you added the soap?

   _____

Experiment 6   119

3. Explain why the soap affected the pin in this manner.

VI. SOLUBILITIES OF COMMON SUBSTANCES IN WATER

1. Data:

	Solute	Water Solubility (S or I)
1	potassium nitrate, KNO_3	
2	sugar, $C_{12}H_{22}O_{11}$	
3	cooking oil, $C_{39}H_{66}O_6$	
4	barium sulfate, $BaSO_4$	
5	isopropyl alcohol, C_3H_8O	
6	1-pentanol, $C_5H_{12}O$	
7	acetic acid, CH_3COOH	

2. Ionic compounds are usually water soluble. List the soluble and the insoluble ionic compounds you tested.

Water-soluble Ionic Compounds	Insoluble Ionic Compounds

3. Molecular compounds that dissolve in water are polar; insoluble molecular compounds are nonpolar. Classify the molecular compounds tested as polar or nonpolar.

Polar Molecular Compounds	Nonpolar Molecular Compounds

Name _____   Section _____   Date _____

Related Questions for Experiment 6

1. When classifying a solution as saturated or unsaturated, why is it important to specify the temperature?

_____

_____

2. What usually happens to the viscosity of a liquid when it is heated?

_____

3. Explain the difference in the boiling points of water (100°C) and seawater (102°C).

_____

_____

4. Explain the difference in the freezing points of water (0°C) and seawater (–8°C).

_____

_____

5. Explain why the boiling point of water averages 94°C in Denver, Colorado, which is 1.6 km (1 mile) above sea level.

_____

_____

6. Pure ethylene glycol (antifreeze) has a freezing point of –11.5°C and water has a freezing point of 0°C. Explain why a 50-50 mixture of ethylene glycol and water remains liquid at –40°C.

_____

_____

7. Explain how the combined use of an ethylene glycol-water antifreeze solution (rather than only water) and a pressurized system can allow automobile cooling systems to reach temperatures of more than 120°C without boiling.

_____

_____

8. If the pressurized radiator cap is removed from a hot car radiator, steam and boiling coolant may suddenly escape and cause severe burns. Explain why the coolant suddenly boils when the radiator cap is removed.

_____

_____

9. Explain why ice melts when rock salt is poured on its surface.

_____

_____

10. Why is a rock salt-ice mixture used in the preparation of home-made ice cream?

_____

_____

Experiment 7

Chemical Reactions

Reference: *General, Organic, and Biological Chemistry: An Integrated Approach, 4th ed., Chapters 5 and 7.*

Purpose: You will perform several reactions to investigate factors that affect the rate of a chemical reaction. Chemical equilibrium and Le Châtelier's Principle will be studied.

Introduction

REACTION RATES

The **rate** of a chemical reaction is the speed at which a reaction occurs. Some reactions, like the rusting of an iron nail, take years to complete. Other reactions, like the explosion of dynamite, are over in a fraction of a second.

In order to react, two molecules must physically collide with each other; the more frequent the collisions, the faster the reaction. The molecules must collide with enough energy (the **activation energy**) to overcome the repelling forces between the molecules. If molecules do not collide with energy at least as large as the activation energy, they will not react. The more collisions meeting or exceeding the activation energy, the faster the reaction.

You will investigate several factors that affect the rate of a chemical reaction: the nature of the reactants, the concentration of the reactants, the surface area of solid reactants, and the temperature. These factors determine the frequency and energy of reactant collisions.

In this experiment you will study the reaction rates of acids with metals under different conditions. You will vary the concentration of reactants by using acids of differing molarity. **Molarity** (M) is defined as the number of moles of solute per liter of solution. If two solutions have equal volumes, the one having a greater molarity will have a greater number of solute molecules.

You will investigate the reaction of hydrochloric acid (HCl) with the metals copper and zinc. The nature of the metals will determine *if* the metal reacts and how fast it reacts. *If* the metals react, they will produce a water-soluble metal chloride along with hydrogen gas. The unbalanced equations are:

$$Cu\ (s)\ +\ HCl\ (aq)\ \longrightarrow\ CuCl_2\ (aq)\ +\ H_2\ (g)$$

$$Zn\ (s)\ +\ HCl\ (aq)\ \longrightarrow\ ZnCl_2\ (aq)\ +\ H_2\ (g)$$

It is possible that the metal will not react:

$$Cu\ (s)\ +\ HCl\ (aq)\ \longrightarrow\ \text{No Reaction}$$

$$Zn\ (s)\ +\ HCl\ (aq)\ \longrightarrow\ \text{No Reaction}$$

In the above equations, (s) represents a solid, (g) a gas, (l) a liquid, and (aq) an aqueous solution.

When a metal reacts with an acid, vigorous bubbling, due to the liberation of hydrogen gas, takes place at the surface of the metal. The presence of hydrogen gas bubbles is evidence that a reaction is occurring. The presence of hydrogen can be verified by placing a burning wooden splint above the surface of reactants. A whistling or popping sound indicates the presence of hydrogen. The lighted splint furnishes the activation energy needed to initiate the reaction between hydrogen and oxygen.

$$\text{hydrogen}\ +\ \text{oxygen}\ \longrightarrow\ \text{water}$$

The intensity of the sound is determined by the concentration of oxygen and hydrogen. Pure hydrogen will produce a muffled "whoof," while the right mixture of hydrogen and air gives a loud "pop." Thus, the intensity of the sound of the reaction between hydrogen and oxygen cannot be used to judge the rate of the reaction between the metal and hydrochloric acid. The rate of reaction between metal and hydrochloric acid can be qualitatively judged by observing how rapidly the metal produces bubbles of gas. A more quantitative determination can be made by timing the number of minutes before the metal disappears.

CAUTION: Most acids are corrosive and can severely damage tissue. See *Procedure* section.

CATALYSIS

A **catalyst** is a substance that increases the rate of a chemical reaction, not by increasing the number or energy of molecular collisions, but by lowering the activation energy required for the reaction. The catalyst interacts with the reactants to produce an alternative path or route for the reaction—a route with a lower activation energy. Reactant molecules are more likely to collide with sufficient energy and the reaction rate is greater in the presence of a catalyst.

The reaction under study in this experiment is the decomposition (breaking apart) of hydrogen peroxide, H_2O_2. Hydrogen peroxide slowly decomposes as it sits in its bottle, gradually turning to water and oxygen gas. The unbalanced equation:

$$H_2O_2\ (aq)\ \longrightarrow\ H_2O\ (l)\ +\ O_2\ (g)$$

This reaction rate is accelerated by light and by catalysts. The presence of oxygen, which indicates that decomposition has taken place, can be easily verified. A glowing wooden splint is slowly burning in the oxygen of air. If the glowing splint is placed in a higher concentration of oxygen, the rate of combustion will increase and the wood splint will burst into flames.

Hydrogen peroxide is an **oxidizing agent**—it removes electrons from other substances. It can destroy organic compounds that come in contact with it. A 3% solution of hydrogen peroxide is used as an antiseptic for small cuts of the skin—it oxidizes germs to death. When placed on an open wound, the hydrogen peroxide bubbles; the bubbles are oxygen gas. The exposed blood contains an enzyme (peroxidase) that catalyzes the decomposition of hydrogen peroxide. The oxygen bubbles come from the hydrogen peroxide, not from the catalyst. A catalyst is neither used up in a reaction nor irreversibly altered; a catalyst is a condition of the reaction and, if included in an equation, is written above the equation arrow. The unbalanced equation:

$$H_2O_2\ (aq)\ \xrightarrow{\text{peroxidase}}\ H_2O\ (l)\ +\ O_2\ (g)$$

When a red spot is suspected of being blood, forensic chemists can use this reaction in a presumptive test for blood. The release of oxygen from hydrogen peroxide by the peroxidase activity causes an indicator to change color. This is not a definitive test for blood since other substances can also catalyze the decomposition of hydrogen peroxide.

Hydrogen peroxide is used in higher concentrations (6%) to bleach (oxidize) hair pigments in many hair-dyeing systems. Still higher concentrations (90%) are sometimes used as oxidizers in rocket propulsion.

CAUTION: Solutions of hydrogen peroxide, especially greater than 3%, should be handled with great care. See *Procedure* section.

CHEMICAL EQUILIBRIUM AND LE CHATELIER'S PRINCIPLE

Some chemical reactions, like the reaction of metal with acid or the decomposition of hydrogen peroxide, "go to completion"—that is, every molecule of reactant is eventually converted to product. Other chemical reactions do not go to completion—the reactant is never totally changed into product. The latter type of reaction is a chemical equilibrium. Chemical equilibriums do not go to completion because of a reverse reaction that competes with the forward reaction. When a reaction is at **equilibrium**, the forward reaction occurs at the same rate as the reverse reaction.

Let's consider a hypothetical chemical reaction in which the reactants are represented by A and B, and the products by C and D. At the beginning of the reaction, only A and B are present, but as the reaction proceeds, C and D are formed. As the concentrations of A and B decrease, the concentrations of C and D increase. The forward reaction may be represented by the equation:

$$A\ +\ B\ \longrightarrow\ C\ +\ D$$

If C or D is not removed from the reaction site, they can react to form the original substances A and B:

$$C \; + \; D \; \longrightarrow \; A \; + \; B$$

As the concentrations of A and B decrease, the forward reaction slows down. As the concentrations of C and D increase the reverse reaction speeds up. Eventually the rate of both reactants becomes equal and an equilibrium is established:

$$A \; + \; B \; \rightleftharpoons \; C \; + \; D$$

In other words, after a period of time, a mixture of products and reactants exists. On the microscopic level, A and B are constantly changing to C and D in the forward reaction and C and D are constantly changing to A and B in the reverse reaction. However, the concentrations of A, B, C, and D remain constant because the forward and reverse reactions occur at the same rate. The composition of this equilibrium mixture will remain constant unless the concentrations of reactants or products are changed by the experimenter. Increasing the concentration of A or B causes the forward reaction to proceed faster, forming additional C and D, until the rate of the reverse reaction increases enough to again establish a state of equilibrium. In the new equilibrium the concentrations of C and D, represented by [C] and [D], are greater than in the original equilibrium.

The nineteenth century French chemist Henri Louis Le Châtelier described this increased concentration of A or B as a "stress." **Le Châtelier's principle** states that if stress is applied to an equilibrium system, the system will respond by altering the equilibrium in such a way as to minimize the stress. Increasing the concentration of A or B is a stress that can be relieved by removing some of the added material and converting A and B to C and D. The equilibrium "shifts to the right" and a new equilibrium is established with higher concentrations of C and D.

Similarly, addition of C or D to a system at equilibrium will cause the equilibrium to "shift to the left" and the new equilibrium position will have higher concentrations of A and B than the original equilibrium. You can also cause equilibrium shifts by removing reactants or products (decreasing their concentrations). Thus, removal of A or B from the equilibrium system will cause a shift to the left. Removal of C or D will cause a shift to the right.

In this experiment you will examine this equilibrium system:

$$FeCl_3 \; (aq) \; + \; 3 \; NH_4CNS \; (aq) \; \rightleftharpoons \; Fe(CNS)_3 \; (aq) \; + \; 3 \; NH_4Cl \; (aq)$$

| iron(III) chloride (pale yellow) | ammonium thiocyanate (colorless) | iron(III) thiocyanate (deep red) | ammonium chloride (colorless) |

Shifts in equilibrium can be easily followed because the color of reactants is different than the color of products. In this equilibrium, when the equilibrium shifts to the right, the mixture becomes more deeply red; when the equilibrium shifts to the left, the mixture becomes less colored. The effect on color intensity (i.e., on the position of equilibrium) will be studied in two ways: (1) by varying the ratio of iron(III) chloride to ammonium thiocyanate and (2) by introducing a salt (ammonium chloride) that contains only the ions not involved in formation of the colored compound, iron(III) thiocyanate.

REACTION OF A PENNY WITH HYDROCHLORIC ACID (Optional)

U.S. pennies made in 1981 or earlier years were made from almost pure copper. Pennies made in 1983 and recent years have a zinc core which is coated with a thin layer of copper. Both types of pennies were made in 1982.

You will investigate one of the newer pennies with a zinc core. By filing away a little of the copper coating, you can expose the zinc core. Which metal will react with hydrochloric acid when the penny is placed in hydrochloric acid? The reaction is too slow to finish in one laboratory period. It is best if you can finish the procedure 24 hours later. The procedure may not work as well if you wait more than 24 hours.

The penny will be destroyed by this procedure. It is not illegal to destroy a penny or any other U.S. legal tender. It is illegal to make a penny look like a dime and attempt to spend it—that is counterfeiting.

Procedure

Since glowing or flaming wood splints are used in this experiment, be sure that no flammable materials are dangerously near.

I. FACTORS THAT AFFECT THE RATE OF A CHEMICAL REACTION

A. *Nature of the Reactants*

> CAUTION: Most acids are corrosive and can severely damage tissue. As always in a chemical laboratory, wear safety goggles while handling acids. If acid is spilled, flood the area with water and notify the instructor immediately. If acid contacts skin, remove any clothing covering the skin and flush with large amounts of water. If acid contacts the eyes, flush with water for 20 minutes and obtain immediate medical attention. Wash your hands after handling acids.

1. Place 5 mL of 6 M hydrochloric acid, HCl, into each of two test tubes. Use **CAUTION** when handling hydrochloric acid. Place the tubes in a test tube rack. Wash your hands after handling HCl.

2. Cut a strip of <u>polished</u> (use emery paper) zinc and a strip of <u>polished</u> copper of equal size, about 1 cm × 1 cm.

3. Place the zinc in one of the test tubes and the copper in the other. Carefully observe each tube. Use a one-hole rubber stopper to <u>loosely</u> stopper each of the tubes for 30 seconds.

4. Remove the stoppers and test for the presence of hydrogen by placing a burning wood splint at the mouth of each of the two test tubes.

5. Clean-up: Discard the liquid of the test tubes into the "Waste Acids" container. Try to keep any metal pieces in the test tube. Transfer the metal pieces to the "Waste Metal" container. Use this same clean-up procedure in Parts B, C, and D.

B. Concentration of the Reactants

1. Pour 5 mL of 6 M hydrochloric acid into one test tube and 5 mL of 1 M hydrochloric acid into a second test tube. Place the tubes in a test tube rack.

2. Cut two pieces of zinc metal, each with a mass of about 0.05 gram, and insert a piece into each of the two test tubes. Note the time of day the zinc is added to the hydrochloric acid. Observe the relative rate of gas bubble formation in each tube.

3. Note the time of day when the reaction ceases (when no zinc remains).

C. Surface Area of a Solid Reactant

1. Weigh approximately 0.05 gram of zinc sheet metal. Place the metal into a clean, dry test tube.

2. Weigh an equal quantity of powdered zinc, and pour it into a second clean, dry test tube.

3. Add exactly 3 mL of 3 M hydrochloric acid to each of the two test tubes. Note the time of day the hydrochloric acid is added to the zinc. Observe each test tube.

4. Note the time of day when the reaction ceases (when no zinc remains).

D. Temperature of the Reactants

1. Prepare an ice-water mixture (bath) in a 250-mL beaker.

2. Pour 5 mL of 6 M hydrochloric acid into each of the test tubes. Place one of the test tubes in the ice-water bath and the other in a test tube rack. Wait 10 minutes for the test tube in the ice bath to cool.

3. Cut two pieces of zinc metal, each with a mass of about 0.05 gram.

4. Simultaneously drop the zinc into each of the two test tubes. Note the time of day the zinc is added to the hydrochloric acid. Observe each test tube. The test tube in the ice bath must remain in the ice bath.

5. Note *for each of the tubes* the time of day when the reaction ceases (when no zinc remains).

II. CATALYSIS

> **CAUTION: Solutions of hydrogen peroxide, especially greater than 3%, should be handled with great care. If spilled, immediately flood the area with water and wash the liquid down the drain. Be careful not to spill any on your person, books, or clothing. Concentrations greater than 3% may cause white patches or blisters and chemical burns on the skin. *Note that a 10% concentration is used here.***

1. Pour 10 mL of 10% hydrogen peroxide (**CAUTION!**) into a test tube. Wash your hands after handling hydrogen peroxide.

2. Loosely stopper the test tube with a one-hole stopper for about 30 seconds.

3. Use a match to light the end of a wooden splint. Allow the fire to burn a few seconds, then blow out the flames, leaving glowing embers.

4. Check for the presence of oxygen by inserting the glowing (not burning) splint deep into the test tube. Observe if the splint bursts into flames. Remove the splint after a few seconds so the test tube does not fill with smoke.

5. Add a small amount (the size of 3 or 4 rice grains) of manganese dioxide to the test tube. Loosely stopper the test tube with a one-hole stopper for about 30 seconds. Note any temperature change. Test for oxygen with a glowing splint.

6. Clean-up: Pour the tube contents into the "Waste Manganese Dioxide" container.

III. CHEMICAL EQUILIBRIUM AND LE CHATELIER'S PRINCIPLE

Iron(III) Chloride–Iron(III) Thiocyanate Equilibrium

1. Note the color of the 0.1 M iron(III) chloride, $FeCl_3$, solution and of the 0.1 M ammonium thiocyanate, NH_4CNS, solution.

2. Pour 50 mL of deionized water into a 150-mL beaker. Add 1 mL of 0.1 M $FeCl_3$ solution and 1 mL of 0.1 M NH_4CNS solution. Stir the mixture; observe the color of the solution.

3. Pour 10 mL of the solution from Step 2 into each of four clean test tubes. Number the test tubes 1 through 4.

4. To test tube No. 1, add 2 mL of deionized water and keep as a reference for color comparison.

5. To test tube No. 2, add 2 mL of 0.1 M $FeCl_3$. Compare the color to test tube No. 1.

6. To test tube No. 3, add 2 mL of 0.1 M NH_4CNS. Compare the color to test tube No. 1.

7. To test tube No. 4, add 2 mL of 0.1 M NH_4Cl. Compare the color to test tube No. 1.

8. Clean-up: Pour the contents of the beaker and test tubes in the "Waste Salts" container.

IV. REACTION OF A PENNY WITH HYDROCHLORIC ACID (Optional)

1. Obtain a U.S. penny with a date of 1983 or more recent. Use a file to remove the copper from the edge of the penny. Either completely remove the copper from the edge or remove the copper from 3 or 4 small places around the edge. Better yet, try one each way.

2. Use a pencil to write your name on the white marking spot of a 100-mL beaker, or use a stick-on label.

3. Place the penny or pennies into the beaker and pour 20 mL of 6 M hydrochloric acid (HCl) into the beaker. Make sure the pennies are covered with acid. Wash your hands.

4. Observe the pennies. Which metal seems to be reacting?

5. Cover the beaker with a watch glass and set it in a safe place for 24 hours.

6. After 24 hours, carefully fill the beaker with tap water to dilute the acid. Dispose of the diluted acid as in Part I.

7. Examine the remains of the pennies very carefully.

Prelab Questions for Experiment 7

_____ 1. The rate of a chemical reaction may be increased by
    a. increasing the frequency of molecular collisions.
    b. increasing the energy of molecular collisions.
    c. lowering the activation energy of the reaction.
    d. all of the above

_____ 2. Increasing the temperature of a reaction
    a. increases the frequency of molecular collisions.
    b. increases the energy of molecular collisions.
    c. increases both the frequency and energy of molecular collisions.
    d. lowers the activation energy of the reaction.

_____ 3. Increasing the concentration of a reactant
    a. increases the frequency of molecular collisions.
    b. increases the energy of molecular collisions.
    c. increases both the frequency and energy of molecular collisions.
    d. lowers the activation energy of the reaction.

_____ 4. Increasing the surface area of a solid reactant
    a. increases the frequency of molecular collisions.
    b. increases the energy of molecular collisions.
    c. increases both the frequency and energy of molecular collisions.
    d. lowers the activation energy of the reaction.

_____ 5. The presence of a reaction catalyst
    a. increases the frequency of molecular collisions.
    b. increases the energy of molecular collisions.
    c. increases both the frequency and energy of molecular collisions.
    d. lowers the activation energy of the reaction.

_____ 6. Which has a greater surface area?
    a. 1 kg of wood sawdust
    b. a piece of solid wood with a mass of 1 kg

7. The rate of a reaction usually doubles for every 10°C increase in temperature. Is heat a catalyst? Explain.

_____

_____

8. What precautions should be observed when handling hydrochloric acid and hydrogen peroxide?

_____

_____

_____

9. Given the following equilibrium, A + B ⇌ C + D, what will be the effect on the equilibrium if each of the following takes place?

	Will equilibrium shift to the left or right?	Will the concentration of these substances increase or decrease?
a. Increase in the concentration of C		A and B
b. Decrease in the concentration of A		C and D
c. Increase in the concentration of B		C and D
d. Decrease in the concentration of D		A and B

Report for Experiment 7

I. FACTORS THAT AFFECT THE RATE OF A CHEMICAL REACTION

A. Nature of the Reactants

    1. Which of the metals reacted with hydrochloric acid?

    _____

    2. Write a complete balanced equation to illustrate the reaction that occurred.

    _____

B. Concentration of the Reactants

    1. Complete the following table:

Solution	Relative Rate of Gas Bubble Formation	Reaction Started (time of day)	Reaction Stopped (time of day)	Reaction Duration (minutes:seconds)
6 M HCl				
1 M HCl				

    2. In which concentration of hydrochloric acid is the reaction faster? _____

    3. What conclusion can you make about the effect of concentration on reaction rate?

    _____

    _____

C. Surface Area of a Solid Reactant

    1. Complete the following data table:

Solution	Relative Rate of Gas Bubble Formation	Reaction Started (time of day)	Reaction Stopped (time of day)	Reaction Duration (minutes:seconds)
Zn Sheet Metal				
Powdered Zn				

2. In which of the zinc/hydrochloric acid solutions was the reaction faster?

_____

3. What conclusion can you make about the effect of surface area on the reaction rate? Was it in accord with your expectation?

_____

_____

D. Temperature of the Reactants

1. Complete the following data table:

Solution	Relative Rate of Gas Bubble Formation	Reaction Started (time of day)	Reaction Stopped (time of day)	Reaction Duration (minutes:seconds)
Ice Water				
Room Temperature				

2. At which temperature was the reaction rate faster?

_____

3. What conclusion can you make about the effect of temperature on reaction rate?

_____

II. CATALYSIS

1. Data. Use (+) for a positive oxygen test and (–) for a negative oxygen test:

a. Hydrogen peroxide _____

b. Hydrogen peroxide with added manganese dioxide _____

2. What function did the manganese dioxide perform?

_____

3. a. Is the decomposition of hydrogen peroxide exothermic or endothermic?

_____

b. How were you able to tell which it was?

_____

c. Write a complete balanced equation illustrating the decomposition of hydrogen peroxide.

_____

d. Draw a potential energy vs. progress of reaction curve in the space below for each of the two systems. Use a solid line for the decomposition of hydrogen peroxide, and a dotted line for the decomposition of hydrogen peroxide in the presence of manganese dioxide.

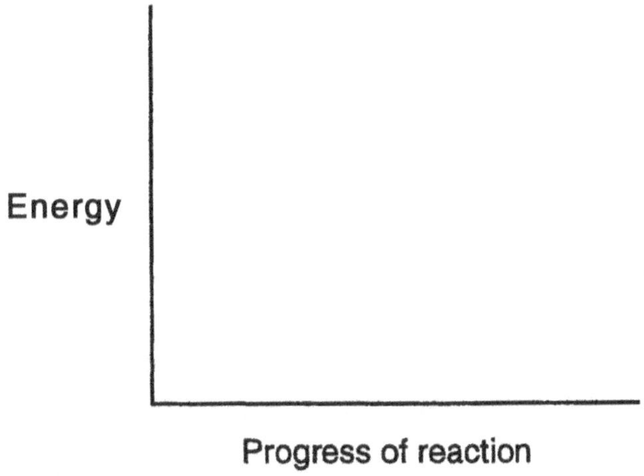

Energy

Progress of reaction

III. CHEMICAL EQUILIBRIUM AND LE CHATELIER'S PRINCIPLE

Iron(III) Chloride–Iron(III) Thiocyanate Equilibrium

1. Data. Color of solutions:

Solution	Color
$FeCl_3$ solution	
NH_4CNS solution	
Mixture of $FeCl_3$ and NH_4CNS solutions	
Test tube No. 1 (added H_2O)	
Test tube No. 2 (added $FeCl_3$)	
Test tube No. 3 (added NH_4CNS)	
Test tube No. 4 (added NH_4Cl)	

2. Write the equation for the iron(III) chloride/iron(III) thiocyanate equilibrium.

_____

3. In which direction did the equilibrium shift, left or right, upon addition of the following reagents?

Solution	Direction of Equilibrium Shift
FeCl$_3$ solution (test tube No. 2)	
NH$_4$CNS solution (test tube No. 3)	
NH$_4$Cl solution (test tube No. 4)	

4. Use the equation for the equilibrium and Le Châtelier's Principle to explain the direction of each shift in equilibrium.

_____

_____

IV. REACTION OF A PENNY WITH HYDROCHLORIC ACID (Optional)

1. Which metal, copper or zinc, reacted with the hydrochloric acid?

_____

2. Write a balanced equation for the reaction that occurred.

_____

3. Describe the remains of the penny after 24 hours in the hydrochloric acid.

_____

Related Questions for Experiment 7

1. Parents have long told their children to thoroughly chew food before swallowing. Use what you know about the factors affecting chemical reaction rates to explain why chewed food is digested faster than unchewed food.

   _____

   _____

   _____

2. Use what you know about the factors affecting the rates of chemical reactions to explain why hospitals take special precautions when oxygen is administered to a patient.

   _____

   _____

   _____

3. Unlike today's blimps that are filled with nonflammable helium, the famous Hindenburg blimp was filled with 190,000,000 liters of hydrogen. After making 10 crossings of the Atlantic Ocean, it was destroyed by fire while landing in New Jersey in 1937, killing 36 of the 92 people on board. The cause of the fire was never determined, but it is likely that static electricity ignited leaking hydrogen. When ignited, hydrogen reacts with oxygen in the air to produce water vapor.

   a. Write a balanced equation for the burning of hydrogen.

   _____

   b. Draw a potential energy curve for the reaction.

Energy

Progress of reaction

c. What is the function of the static electricity?

4. Commercial hydrogen peroxide is stored in dark bottles. Explain why.

_____

_____

5. Give an example of stress on a chemical equilibrium in your body. Refer to your text.

_____

_____

_____

Experiment **8**

Stoichiometry: Mole Relationships

Reference: *General, Organic, and Biological Chemistry: An Integrated Approach, 4th ed., Chapters 3, 5, and 7.*

Purpose: You will establish the validity of a balanced equation by experimentally determining the molar ratio of reactants and products in a chemical reaction. From this you will distinguish between two possible starting materials.

Introduction

REACTIONS AND EQUATIONS

Propane (C_3H_8) is a gaseous fuel sold in steel cylinders and used in camping stoves, propane torches, and gas barbecue grills. The combustion of propane is an example of a chemical reaction. A chemical reaction and a chemical equation are not the same. You cannot see this reaction until you have lit your camp stove. A chemical equation is a symbolic representation of the chemical reaction. The equation for the reaction is:

$$C_3H_8 \; + \; 5\,O_2 \longrightarrow 3\,CO_2 + 4\,H_2O$$

Chemical reactions are automatically balanced by nature—mass is neither created nor destroyed in a chemical reaction. Chemical equations are not automatically balanced. A chemist must balance chemical equations by inserting coefficients in front of the reactants and products.

The coefficients of a balanced equation can be interpreted in two ways:

1. **Coefficients as number of molecules.** The equation above can be read as:

   "one molecule of propane reacts with five molecules of oxygen to produce three molecules of carbon dioxide and four molecules of water."

2. **Coefficients as number of moles.** Since molecules are so tiny and since so many molecules are involved in a reaction, chemists count molecules in a larger unit called a mole. One mole of any substance is 6.02×10^{23} molecules. Therefore the equation can also be read as:

   "one mole of propane reacts with five moles of oxygen to produce three moles of carbon dioxide and four moles of water."

The actual number of moles of propane that burn in the camp stove depends on the length of time the stove is lit. No matter how long the stove is lit, the ratio of moles of oxygen consumed to moles of propane consumed will remain at 5:1. **The coefficients of the equation represent the simplest whole number ratios of moles of reactants and products.**

BALANCING EQUATIONS BY TRIAL-AND-ERROR

Aluminum metal reacts with hydrochloric acid to produce aluminum chloride and hydrogen gas. The unbalanced equation can be written as:

$$Al + HCl \longrightarrow AlCl_3 + H_2$$

The equation can be balanced by trial-and-error. Usually it is best to balance hydrogen and oxygen last, after all other elements have been balanced. Let's begin by noticing that aluminum is already balanced. Balance chlorine by adding a coefficient of 3 HCl.

$$Al + 3\,HCl \longrightarrow AlCl_3 + H_2$$

There are three hydrogen atoms on the left and two on the right side. We need six hydrogen atoms on each side to balance the equation.

$$Al + 6\,HCl \longrightarrow AlCl_3 + 3\,H_2$$

The chlorine atoms have become unbalanced. The left side has 6 chlorine atoms and the right side has 3 chlorine atoms. To balance chlorine place a coefficient of 2 in front of $AlCl_3$.

$$2\,Al + 6\,HCl \longrightarrow 2\,AlCl_3 + 3\,H_2$$

The final step is balanced since there are two aluminum, six hydrogen, and six chlorine atoms on each side of the equation. Mass is conserved; no atoms are created or destroyed in the chemical reaction.

BALANCING EQUATIONS FROM EXPERIMENTAL DATA

We could experimentally verify the coefficients of the equation by reacting two moles of aluminum with six moles of hydrochloric acid. If the products were collected and measured there would be two moles of aluminum chloride and three moles of hydrogen produced in the reaction.

Actually we can begin with any amount of aluminum and still demonstrate the validity of the equation. Consider the following data collected by a student. She measured the masses of aluminum and hydrochloric acid that were used in her experiment and the masses of aluminum chloride and hydrogen that were produced.

Substance	Mass
Al	103 g
HCl	416 g
$AlCl_3$	507 g
H_2	11.5 g

Let's see how her experimental data verifies the validity of the balanced equation.

Step 1: Calculate the number of moles of each substance. The formula weight of a substance can be found by adding the atomic weights of every atom in the formula.

Formula weight of Al = 26.98 amu

Molecular weight of HCl = 1.01 + 35.45 = 36.46 amu

Formula weight of $AlCl_3$ = 26.98 + (3 × 35.45) = 133.33 amu

Molecular weight of H_2 = 2 × 1.01 = 2.02 amu

The mass of 1 mole of a substance is equal to the formula weight or molecular weight expressed in units of grams. The mass in grams of one mole (the molar mass) can be used as a conversion factor. [It hardly seems worthwhile to abbreviate a four-letter word, but because "mole" is used so often, it is commonly shortened to "mol."]

(103 g Al) (1 mol Al/26.98 g Al) = 3.82 mol Al

(413 g HCl) (1 mol HCl/36.46 g HCl) = 11.4 mol HCl

(507 g $AlCl_3$) (1 mol/133.33 g $AlCl_3$) = 3.80 mol $AlCl_3$

(11.5 g H_2) (1 mol H_2/2.02 g H_2) = 5.69 mol H_2

The coefficients of an equation represent the number of moles of each substance. Therefore we might try to write the equation for the reaction as:

$$3.82 \text{ Al} + 11.4 \text{ HCl} \longrightarrow 3.80 \text{ AlCl}_3 + 5.69 \text{ H}_2$$

This is incorrectly written since **the coefficients must be the smallest whole number molar ratios of the substances.**

Step 2: Find the molar ratios. To find the ratios of moles of the reactants and products, divide each number by the smallest number. The smallest number is 3.80.

Substance	Molar ratios
Al	3.82 ÷ 3.80 = 1.01
HCl	11.4 ÷ 3.80 = 3.00
$AlCl_3$	3.80 ÷ 3.80 = 1.00
H_2	5.69 ÷ 3.80 = 1.50

We could use the molar ratios to write the equation as:

$$1.01 \text{ Al} + 3.00 \text{ HCl} \longrightarrow 1.00 \text{ AlCl}_3 + 1.50 \text{ H}_2$$

This is still incorrect since the coefficients of the equation must be the smallest whole number ratios. The coefficient 1.01 Al is very close to a whole number but the coefficient 1.50 H_2 is not close to a whole number.

Step 3: Find the whole number molar ratios. If the molar ratios are not whole numbers, multiply each coefficient by 2 to obtain whole numbers. If 2 does not produce whole numbers, continue using the next higher integer (3, 4, 5) until roundable to a whole number.

Substance	Mass	Number of moles	Molar Ratios	Whole Number Molar Ratios
Al	103 g	3.82 mol	1.01	2
HCl	416 g	11.4 mol	3.00	6
$AlCl_3$	507 g	3.80 mol	1.00	2
H_2	11.5 g	5.69 mol	1.50	3

These whole number molar ratios are based on experimental data. Since measurements are never exact, the calculated whole numbers are not exactly whole numbers. We had to round off 2.02 to 2 and use it as the coefficient of aluminum.

The whole number molar ratios are the coefficients of the balanced equation!

$$2\ Al\ +\ 6\ HCl\ \longrightarrow\ 2\ AlCl_3\ +\ 3\ H_2$$

Two moles of aluminum react with six moles of hydrochloric acid to produce two moles of aluminum chloride and three moles of hydrogen. We have experimentally determined the coefficients of the equation by measuring the masses of each reactant and product.

Look again at the masses of the reactants and products.

Mass of reactants = 103 g + 416 g = 519 g
Mass of products = 507 g + 11.5 g = 518.5 g

The Law of Conservation of Mass states that in a chemical reaction, the total mass of the reactants equals the total mass of the products. Since experimental measurements are never exact, small differences in the masses occur but are not significant.

Summary: To find the smallest whole number coefficients of a balanced equation using experimental data:

1. **Find the number of moles of each substance.**

2. **Find the molar ratios by dividing each number of moles by the smallest number of moles.**

3. **If the molar ratios are not whole numbers, multiply the molar ratios by an integer to obtain whole numbers.**

DIFFERENTIATION BETWEEN TWO CARBONATES, OBJECT AND PLAN

Your instructor will assign you an unknown, identified only by number. It will be either **sodium hydrogen carbonate** ($NaHCO_3$; in cooking it's called baking soda) or **sodium carbonate** (Na_2CO_3). Its decahydrate, $Na_2CO_3 \cdot 10H_2O$, is called washing soda, but you will use the carefully dried anhydrous form. You are then to react a weighed sample of your unknown with a small excess of hydrochloric acid, which will produce sodium chloride, carbon dioxide and water as shown in these unbalanced equations:

$$NaHCO_3 \ + \ HCl \ \longrightarrow \ NaCl \ + \ CO_2 \ + \ H_2O$$

$$Na_2CO_3 \ + \ HCl \ \longrightarrow \ NaCl \ + \ CO_2 \ + \ H_2O$$

The CO_2, being a gas, conveniently escapes. Then, by evaporating the water and excess hydrochloric acid, you can isolate and weigh the NaCl obtained from a known weight of unknown. To get an accurate weight on the NaCl produced, *it will be important not to lose any of the sample during its reaction with HCl and the subsequent evaporation of the water and excess HCl. At the same time, it will be important to be sure your NaCl is completely dry so a second heating and weighing is required.*

When you have determined the weight of dry NaCl formed, you can determine which of the two possible carbonates is your unknown by determining the mole ratio of NaCl product to starting carbonate. The mole ratio for *one* of the two carbonates, the one you have in your unknown, will be a small whole number ratio, 1 to 1 or 2 to 1. This will also provide experimental evidence from which you can balance the reaction equation.

Procedure

Record all of your data in the table provided below. *Obtain all weights (masses) accurately to two decimal places and be sure to specify all units that are involved.* The evaporating dish and watch glass cover used must be cooled to room temperature before weighing, otherwise convection currents can make the balance readings unreliable.

UNKNOWN NUMBER __11__

Weigh a **dry** porcelain evaporating dish and then weigh into it an approximate (but carefully weighed) 4 g sample of your unknown. For example, your sample might weigh something like 3.86 g or 4.12 g; it doesn't have to be exactly 4.00 g.

Mass of unknown plus evaporating dish   _____

Mass of evaporating dish   _____

Mass of the unknown sample   _____

Then slowly add about 1-mL portions of 3 *M* hydrochloric acid until all of your sample has dissolved and no more bubbles of carbon dioxide are being formed. DO NOT splash out any contents of the dish; the object is to obtain every bit of the NaCl formed. You can stir the mixture, but the stirring rod must not be taken out of the mixture without rinsing it with a few drops of the hydrochloric acid. To complete the reaction of your carbonate sample will require from 20 to 25 mL of the 3 M HCl.

You can check to see that you have an excess of acid by stirring the reaction mixture and touching the stirring rod on a strip of blue litmus paper; it will change color to "red" (pink) if the mixture is acidic. Do NOT dip litmus paper into the reaction mixture.

Weigh a watch glass (preferably the ribbed type) that will be used to cover the evaporating dish:

Mass of the watch glass to be used   _____

Place the covered dish on a hot plate, and heat it so that the contents boil **gently**. Because you want to avoid loss of product, try to keep the spattering to a minimum by keeping the watch glass in place and controlling the rate of heating. Keep the watch glass on the evaporating dish throughout the heating and weighing (the second time also) because some NaCl probably will have spattered up onto it.

When the residue in the dish appears to be completely dry, *use crucible tongs* to remove the dish and watch glass from the hot plate and set them on an iron ring attached to a ring stand. After the covered dish cools to room temperature, determine the weight of the dish, watch glass, and residual NaCl. Subtract the combined weight of the empty evaporating dish and watch glass to get the weight of the NaCl produced. To make sure that all water has been evaporated, reheat the dish and then recool and reweigh it.

Mass of NaCl plus evaporating dish & watch glass *after initial heating*   _____

Mass of NaCl plus evaporating dish & watch glass *after second heating*   _____

Combined mass of evaporating dish & watch glass   _____

Mass of the NaCl obtained   _____

Now use the unknown and NaCl weights to calculate their mole ratio (see **Report** section).

Discard the NaCl residues by washing them into a sink drain.

Prelab Questions for Experiment 8

1. Why must all masses be determined at room temperature?

    _____

    _____

2. The following equation represents a commonly used method for producing oxygen. Balance it by the trial-and-error method.

$$KClO_3 \longrightarrow KCl + O_2$$

3. When 6.50 g of a red mercury oxide (X) were heated, the compound decomposed into mercury and oxygen.

$$\text{Compound X} \longrightarrow Hg + O_2$$

Using the data below, calculate the ratio of mercury and oxygen atoms in the compound (this is called its empirical formula) and balance the equation.

	Mass	Number of Moles	Molar Ratio	Whole Number Molar Ratio	Ratio of Atoms
Hg	6.04 g	_____	_____	_____	_____
O_2	0.47 g	_____	_____	_____	_____

Calculations:

a. The formula of compound X is:

b. The balanced equation for its reaction on heating is:

c. The name of compound X is:

Report for Experiment 8

IDENTIFYING THE UNKNOWN

Recall, the number of moles = mass/molar mass, so here moles of NaCl = mass NaCl/(58.5 g/mol).

Similarly, **if** the unknown were $NaHCO_3$ (call it compound A), the sample would contain this many moles: sample weight/(84.0 g/mol) AND **if** it were Na_2CO_3 (call it compound B), it would contain this many moles: sample weight/(106.0 g/mol).

Make these calculations and enter your values in this table:

Possible Starting Material	Molar Mass	Number of Moles IF Unknown Is	Mass (g) of NaCl Obtained	Number of Moles of NaCl (C)	Mole Ratios	Whole Number Mole Ratio*
$NaHCO_3$	84.0	A:			C/A =	
OR						
Na_2CO_3	106.0	B:			C/B =	

*This is the ratio of moles of NaCl produced to moles of carbonate used.

Remember that the whole number mole ratio for *one* of the two carbonates, the one you have in your unknown, will be close to 1 to 1 **or** 2 to 1.

RESULTS

1. Your unknown is which carbonate?

2. Using your conclusion on unknown identity and whole number mole ratios, write the balanced equation for the reaction of your unknown with hydrochloric acid.

3. What specific experimental errors might cause your whole number mole ratios to differ from the coefficients of the balanced equation obtained by the trial-and-error method?

4. Considering the products of the reaction, what (small) climatic effect has your experiment had?

Stoichiometry: Mole Relationships

Related Questions for Experiment 8

Butane, C_4H_{10}, is a common fuel found in cigarette lighters. When butane burns, it combines with oxygen to produce carbon dioxide and water. A student burned a 5.00-gram sample of butane. Show all calculations in solving the following problems.

1. Write a balanced equation for the combustion of butane.

2. Find the number of moles of butane in 5.00 grams.

3. Find the number of moles of oxygen that will react with 5.00 g of butane.

4. Find the number of grams of oxygen that will react with the butane.

5. Find the number of moles of carbon dioxide produced.

6. Find the number of grams of carbon dioxide produced.

7. Find the number of moles of water produced.

8. Find the number of grams of water produced.

9. Find the total mass of the reactants and compare it to the total mass of the products.

Properties of Gases

Reference: *General, Organic, and Biological Chemistry: An Integrated Approach, 4th ed., Chapter 6.*

Purpose: You will prepare carbon dioxide and observe its key properties. You will study absolute
temperature and the behavior of gases under changing conditions of pressure and temperature.

Introduction

CARBON DIOXIDE GAS: PREPARATION AND PROPERTIES

Carbon dioxide gas is one of the endproducts of digesting food and is in the air we exhale, which
is about 4% CO_2. The atmosphere we breathe contains only 0.03% CO_2. The presence of dissolved
CO_2 in the blood stimulates breathing. For this reason, CO_2 is added to the oxygen gas used to
assist respiration and to the gases used in anesthesia.

Carbon dioxide is also produced when an organic (carbon-containing) fuel burns. Fuels such as
gasoline, coal, natural gas, and wood all produce CO_2 when they burn, increasing their
concentration in our atmosphere. Carbon dioxide is a "greenhouse gas," one that efficiently absorbs
heat from the sun and raises the temperature of the atmosphere; i.e., it contributes to global
warming. Other important greenhouse gases are methane (CH_4) and nitrous oxide (N_2O), which is
also called "laughing gas." All three are naturally occurring gases, but their concentrations are
being increased by modern rates of industrial and agricultural production.

Plants remove CO_2 from the air via photosynthesis, the solar-powered conversion of CO_2 to
carbohydrates. Carbon dioxide is also removed from the air by dissolving in seawater, slightly
increasing its acidity because of this equilibrium:

$$CO_2 + H_2O \rightleftharpoons H_2CO_3 \rightleftharpoons H^+ + HCO_3^-$$

$$\text{carbonic acid} \qquad \text{hydrogen} \quad \text{bicarbonate}$$
$$\text{ion} \qquad \text{ion}$$

Carbonated beverages ("sodas") contain much higher concentrations of CO_2 because they are
bottled under high CO_2 pressure (see Henry's Law below). It is the above equilibrium that explains
their characteristic acidity and tartness.

In baking cookies and cakes, a chemical reaction produces CO_2 from baking soda, $NaHCO_3$, when
it reacts with acids like lactic acid, found in skim and sour milk.

$$NaHCO_3\ (s) + \underset{\text{lactic acid}}{CH_3\overset{HO}{\underset{|}{C}}H\overset{O}{\overset{||}{C}}OH}\ (aq) \longrightarrow CO_2\ (g) + H_2O\ (l) + \underset{\text{sodium lactate}}{CH_3\overset{HO}{\underset{|}{C}}H\overset{O}{\overset{||}{C}}ONa}\ (aq)$$

[The parenthetical letters mean: *s* for solid, *l* for liquid, *g* for gas, and *aq* for water (aqueous)
solution.]

Some foods contain acidic material that suffices for the reaction. When they do not, baking soda must be used; it is a mixture of $NaHCO_3$ and an acid that, only after being put into a moist batter, can interact and produce CO_2. The usual acidic component of baking powder is the mono potassium salt of the diacid tartaric acid, a byproduct of wine making. To make bread and pizza dough rise, the needed CO_2 is produced by yeast acting on sugar.

Carbon dioxide is not only nonflammable, it can smother a fire. It does this by displacing the air around the burning object. Since the oxygen in the air is necessary for combustion, the fire goes out. The relative density of CO_2 and air, which can quickly be calculated as explained below, is a further reason for its effectiveness.

Although carbon dioxide is a gas at room temperature, it is a solid at temperatures below $-78.5°C$ ($-109.3°F$). Solid carbon dioxide is called "dry ice" since it changes from solid to gas without ever forming a liquid, a process called **sublimation**.

$$CO_2\ (s) \xrightarrow{\text{heat}} CO_2\ (g)$$

Manufacturing carbon dioxide is big business. Billions of tons of carbon dioxide are produced each year in the United States making it a commercially important material.

> **CAUTION! Do not touch dry ice with your bare skin; the extreme cold can severely damage the skin. If used in an enclosed space, there is danger of asphyxiation by the gaseous carbon dioxide produced.**

THE GAS LAWS

Not all gases are colorless, odorless, solidify at $-78.5°C$, and extinguish fires; these are properties of carbon dioxide gas. All gases do share certain properties described by the gas laws if the gases are at normal or high temperature and normal or low pressure, the conditions under which their component molecules are in random motion.

BOYLE'S LAW: THE PRESSURE-VOLUME LAW

Boyle's Law states that the volume of a gas is *inversely* proportional to the pressure when the temperature remains constant. That is, the volume of a gas sample increases as the pressure decreases; conversely, the volume decreases as the pressure on it increases.

In the experiment that qualitatively illustrates Boyle's Law, a water aspirator is used to decrease the pressure in the equipment. It is also the usual source of the vacuum needed for vacuum filtration (see Fundamental Laboratory Operations, Part 8, p. 24). An aspirator is attached to a cold-water faucet. When water flows rapidly through it, suction is created in the side arm, a result predicted by Bernoulli's Principle. It states that as the velocity of a liquid or gas is increased, as it is when the water is forced through the restriction inside the aspirator (see Figure 9-1), its pressure is reduced. This is counterintuitive, but it works to draw a gas (or liquid) from equipment attached to the aspirator's side arm. It is the same principle that explains the operation of spray bottle attachments connected to a garden hose for spraying herbicides or window washing solutions.

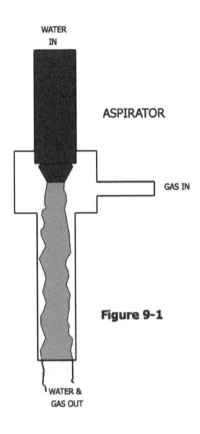

WATER
IN

ASPIRATOR

GAS IN

Figure 9-1

WATER &
GAS OUT

A water aspirator cannot reduce pressure to lower than the vapor pressure of the flowing water, which increases as its temperature increases. The lowest pressure of cold water is about 10-20 mmHg. For comparison, atmospheric pressure averages 760 mmHg (this is the height of the mercury column in a mercurial barometer) at sea level and decreases with increasing altitude.

For demonstrating Boyle's Law, a small, sealed balloon is placed into a filter flask. See Figure 9-2. A water aspirator is used to reduce the gas (air) pressure in the filter flask and, indirectly, in the sealed balloon. It expands until the gas pressure inside the balloon equals that in the flask.

In the middle of the Figure 9-2 diagram there is a second filter flask, midway between the aspirator and the equipment being used.

This trap is always important when using an aspirator. It stops water from being sucked into the experimental equipment if you turn off the aspirator water before disconnecting the hose between it and the aspirator. It also protects the aspirator by stopping liquids from being sucked into it from attached equipment, e.g., if too much filtrate gets into a vacuum filtration receiver flask.

The physiological application of Boyle's Law can be illustrated with the equipment shown in Figure 9-3, a lung demonstration model. It shows the role of the diaphragm in drawing air in and out of the lungs.

ABSOLUTE TEMPERATURE

Temperature is a measure of the average kinetic energy of the particles in a sample of matter. At room temperature the particles are moving very rapidly. When temperature is decreased, the particles have lower kinetic energy because they slow down. Theoretically, at some very low temperature the particles would become motionless, i.e., their kinetic energy would reach zero. This temperature, called absolute zero, is equal to -273 °C or, on the absolute temperature scale, 0 K (for Kelvin, the physicist who defined the scale). Modern theory says that absolute zero can never be reached; that particle motion can never completely cease. Nevertheless, temperatures have been reached that very nearly reach absolute zero.

To convert from temperature on the Celsius scale (°C) to absolute temperature (K), simply add 273 to the numerical value. For example room temperature, usually about 25 °C, becomes 298 K.

CHARLES' LAW: THE TEMPERATURE-VOLUME LAW

Charles' Law states that the volume of a sample of gas varies *directly* with the absolute temperature if the pressure is not changed. That is, the volume of a gas sample will increase when the temperature increases (and vice versa) if the pressure is kept constant. For example, a 200-mL sample of air at room temperature 25 °C (298 K) on 75 °C increase in temperature to 100 °C (373 K) by heating in boiling water, would increase in volume to (200 mL) (373 K/298 K) = 250 mL (*if its pressure is held constant*).

GAY-LUSSAC'S LAW: THE PRESSURE-TEMPERATURE LAW

In a rigid container (one whose volume remains constant) that holds a gas sample, increasing the temperature will increase the gas pressure. Lowering the temperature will lower the pressure. Gay-Lussac's Law states that the pressure of a constant-volume gas sample varies in direct proportion to the absolute temperature.

For example, an auto tire maintains an essentially constant volume even as its temperature changes, but the tire pressure increases as its temperature increases. This explains why it is important to check tire pressure when tires are cool. To illustrate, an initial temperature of 298 K that increases in use by 90 K will have its pressure increase from an initial 32 psi (pounds per square inch) to:

(32 psi)(388 K/298 K) = 42 psi, a considerably higher pressure.

AVOGADRO'S LAW

This law says that at constant pressure and temperature, the volume of a gas is directly proportional to the amount of gas in the sample. A convenient mathematical expression of this law is:

> At standard temperature (which is defined as 0 °C, 273 K) and standard pressure (which is defined as 1 atmosphere), the volume of 1 mole **of any gas** is 22.4 L. [One atmosphere is the pressure of the air around us when at sea level; it equals 760 mmHg or 14.5 psi.]

One recurring use of this relationship is that it enables quick calculation of the relative density of different gases. For example, what is the density of methane, the principal ingredient of natural gas? To minimize the risk of it leaking from a gas furnace, will it be better to provide escape vents above or below the furnace?

Since 1 mole of each gas will occupy 22.4 L at standard temperature and pressure, methane's density relative to air (principally nitrogen) is:

$$\frac{(16 \text{ g of methane/mol})/22.4 \text{ L}}{(28 \text{ g nitrogen/mol})/22.4 \text{ L}} = 0.57$$

So methane is much less dense than air and can best escape through high vents.

THE IDEAL GAS LAW

The individual gas laws above can all be combined into the easy to use formula:

PV = nRT.

Here, **P** means pressure, **V** means volume, **n** means number of moles of the gas, **T** means absolute temperature, and **R** is a constant whose numerical value depends on the units used for the other terms. Using the most frequently used units, R has the value:

R = PV/nT = [(1 atmosphere pressure)(22.4 liters volume)]/ (1 mole)(273 K)
= 0.0821 L-atm/mole K

SOLUBILITY OF GASES: HENRY'S LAW, THE PRESSURE-SOLUBILITY LAW, AND SOLUBILITY EFFECT OF TEMPERATURE

According to Henry's Law, the solubility of a gas is *directly* proportional to the pressure of the gas on the liquid. Increasing the pressure of the gas increases its solubility and vice versa. A carbonated beverage container is sealed while it is under a high pressure of carbon dioxide, which results in its having a high concentration of dissolved CO_2. While dissolved, the gas is invisible—the mixture is homogeneous. Opening the container decreases the pressure and results in bubbles of CO_2 forming and rising to the surface.

Temperature also has an effect on gas solubility, an effect that is the opposite of the usual effect on the solubility of solids in a liquid. The solubility of gases *decreases* as the temperature goes up. An example of the significance of this behavior is in the adverse-for-fish decreased concentration of dissolved oxygen in water as a result of thermal pollution from use of river water to cool nuclear reactors.

Procedure

I. CARBON DIOXIDE GAS: PREPARATION AND PROPERTIES

1. Make a candle holder by cutting a piece of corrugated cardboard to a size of about 5 cm × 5 cm. Use a sharp awl to pierce a small hole in the center of the cardboard.

2. Insert a birthday candle through the hole. The candle should stand upright on the cardboard base.

3. Place the cardboard and candle in the bottom of a 400-mL beaker. Then cover the bottom of the beaker with 10 g of baking soda (sodium bicarbonate, $NaHCO_3$).

4. Obtain about 30 mL of vinegar (5% acetic acid, CH_3CO_2H) in a 50-mL beaker.

5. Check to be sure no flammable materials are nearby! Holding a lighted match with tongs, light the candle.

6. All at once, pour the vinegar down the inside edge of the beaker, NOT ON THE CANDLE! Observe what happens. If the candle goes out, try to re-light it.

II. DRY ICE

1. Using tongs or insulated gloves, place a small piece (about the size of a marshmallow) of dry ice in a quart-size zip-close plastic bag. **CAUTION: Do not touch the dry ice with your bare skin; the extreme cold can cause serious injury.**

2. Flatten the bag to remove as much air as possible and seal the bag shut.

3. Allow the dry ice to slowly warm up. Observe.

III. BOYLE'S LAW: THE PRESSURE-VOLUME LAW

A. Volume-Pressure Relationship

1. Slightly inflate a small balloon to a size so that it can be placed inside a 250-mL suction flask. Knot the end tightly to prevent air from escaping.

2. Set up the apparatus as illustrated in Figure 9-2 but do not insert the stopper. Securely attach one length of rubber vacuum tubing from the stem of the flask to the trap. The second length of vacuum tubing connects the trap to the side arm of the water aspirator. The vacuum tubing has thick walls to prevent its collapse.

3. Run the water at full force and place the palm of your hand on top of the suction flask. If the aspirator is working properly you should feel a strong suction due to the low pressure inside the suction flask.

4. Without turning off the water, remove your hand and tightly insert the stopper into the suction flask. Observe the behavior of the balloon in the flask.

5. Detach the vacuum tubing from the aspirator and allow air to enter the flask. Observe the behavior of the balloon.

6. Shut off the water. *Save the suction flask set-up for Part V.*

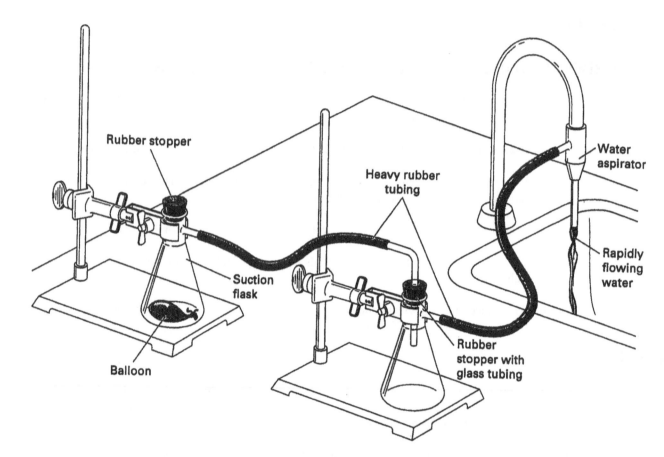

Figure 9-2 The setup for studying volume-pressure relationship.

B. Physiological Application of Boyle's Law

1. Observe a lung demonstration model like that shown in Figure 9-3. If one is unavailable, simply think through how it would perform in the below operations. The upper end of the Y-tube is open to the air, allowing air to enter and leave the balloons. A thin sheet of rubber is stretched across the bottom of the bell jar. The rubber stopper holding the Y-tube and the

rubber sheet seal the bell jar airtight. No new air can enter or leave the bell jar, but the air already inside can change in volume and pressure as the sheet rubber diaphragm is alternately pulled out and released.

2. Pull out the rubber diaphragm as illustrated in Figure 9-3. Observe the behavior of the balloons.

3. Release the rubber diaphragm, allowing it to return to its original position. Observe the behavior of the balloons.

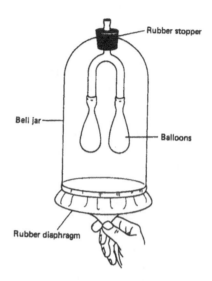

Figure 9-3 Lung demonstration model.

IV. CHARLES LAW: THE TEMPERATURE-VOLUME LAW

See Figure 9-4. Obtain (or assemble) one of the balloon-on-glass tubing assemblies. Place about 20 mL of room-temperature water in the 100-mL graduated cylinder (it need not be one that can be stoppered) and then carefully place the tubing assembly (with balloon **uninflated** and pinch clamp open) into the cylinder. [The bottom of the tube should be about half an inch up from the bottom of the cylinder.] If the balloon appears to be slightly inflated, use an ear syringe or pipet bulb to draw air out. The water level should be just *slightly* above the short length of thick-walled vacuum tubing, which is used just to make a tighter seal of the balloon against the glass tubing. If you have to add more water, *add the minimum needed* and record in the Report section's table (which has spaces for recording volumes A through E and your calculations) the combined volume of water and the submerged part of the tubing assembly (**volume A**).

Then, using an ear syringe or pipet bulb, blow just enough air into the balloon to increase the combined volume of balloon assembly, water and air to about 80 mL and *tighten* the pinch clamp.

Now, immerse the graduated cylinder to at least ¾ of its height in a 1- or 2-L beaker containing ice. When, in 10 or 15 minutes, its volume reading has reduced to a fairly stable value, observe and record the water level in the graduated cylinder (**volume B**). Volume B minus volume A equals the volume of air in the balloon at 0 °C, **volume C**.

Then, place the graduated cylinder in a warm water bath, one at about 75-80 °C. After about 10 or 15 minutes, when the volume of the cylinder contents is no longer changing, record the water level in the graduated cylinder (**volume D**) and *record the exact temperature*. Calculate the air volume (**E**) at this temperature (**volume E** = volume D minus volume A).

Using Charles' Law (gas volume varies directly with the absolute temperature), calculate what the volume of the 0 °C volume of air used in the experiment (volume **C**) should theoretically be at your elevated temperature. Show your calculations and discussion of *observed volume* **E** versus your *calculated volume* **E** based on Charles' Law in the Report section. If there is a discrepancy, suggest how this could be explained.

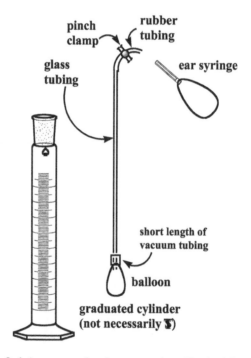

Figure 9-4 Apparatus for demonstrating Charles' Law

V. HENRY'S LAW: THE PRESSURE-SOLUBILITY LAW

1. Remove the balloon used in Procedure III.A. from the suction flask.
   Empty any water that may have entered the trap.

2. Pour 25 mL of carbonated water into the suction flask.

3. Connect the vacuum tubing to the water aspirator.

4. Run the water at full force and observe the behavior of the carbonated water.

5. Disconnect the vacuum tubing and observe the behavior of the carbonated water.
   Shut off the water.

VI. GAY-LUSSAC'S LAW: THE PRESSURE-TEMPERATURE LAW AND ABSOLUTE ZERO (Instructor's Optional Demonstration)

In this demonstration, the equipment needed is basically just a pressure gauge connected to a hollow metal ball that contains dry air. (See, e.g., item number WL1077D on the Sargent-Welch website.) The ball will be immersed in liquids at three different temperatures and the pressure measured in each case. These will be plotted on graph paper versus Celsius temperature. Then a best-fitting straight line will be drawn.

1. Your instructor will immerse the bulb of the apparatus in three different liquid baths at three different temperatures. The recommended ones are boiling water (100 °C), ice water (0 °C), and dry ice-isopropyl alcohol (-78.5 °C). Record the pressure and temperature for each of the liquids.

2. Plot the three data points on the graph in the Report section (see Figure 9-5). Note that pressure is on the vertical axis and temperature on the horizontal axis.

3. Use a ruler and draw one single straight line that comes closest to passing through all three data points.

4. Extend the straight line to lower pressures and temperatures. Determine the point at which the pressure is zero. This temperature is absolute zero. It should be at -273 °C, but there may be errors in pressure measurement. Also, more accuracy would be obtained if some additional, lower temperatures could be used.

Properties of Gases

Prelab Questions for Experiment 9

1. What safety precautions are necessary when handling dry ice?

   _____

   _____

2. Describe the order of steps in shutting off a water aspirator.

   _____

   _____

3. How does the volume of a gas change when the pressure is increased?

   _____

   _____

4. How does the volume of a gas change when the temperature is increased?

   _____

   _____

5. How does the water-solubility of a gas change when the pressure is increased?

   _____

   _____

6. In Part III. B., which parts of the human body are represented by the following pieces of the lung demonstration model:

   a. The two rubber balloons _____

   b. The Y-tube _____

   c. The open end of the Y-tube _____

   d. The bell jar _____

   e. The thin sheet of rubber _____

Properties of Gases

Report for Experiment 9

I. CARBON DIOXIDE GAS: PREPARATION AND PROPERTIES

1. What is the evidence that baking soda and vinegar react to produce a gas?

   _____

2. What properties does carbon dioxide gas have (color, effect on fires, etc.)?

   _____

   _____

3. Were you able to relight the candle? _____

4. Is there any evidence that carbon dioxide gas is lighter or heavier than air? If so, what?

   _____

   _____

5. Using the method illustrated in the *Introduction* (see Avogadro's Law), calculate the relative density of CO_2 and air. Show the calculation.

II. DRY ICE

1. What happened to the dry ice in the bag?

   _____

2. How could you test the gas in the bag to show it was carbon dioxide?

   _____

   _____

III. BOYLE'S LAW: THE PRESSURE-VOLUME LAW

A. Volume-Pressure Relationship

1. a. What happened to the balloon when you decreased the pressure in the suction flask?

   _____

b. Explain your observations in Question 1a.

_____

_____

2. a. What happened to the balloon when you detached the vacuum tubing from the aspirator?

_____

b. Explain your observations in Question 2a.

_____

_____

B. Physiological Application of Boyle's Law

1. a. What happened to the balloons when you pulled out the rubber diaphragm?

_____

b. Explain your observations in *a*.

_____

_____

2. a. What happened to the balloons when the diaphragm returned to its original position?

_____

b. Explain your observations in *a*.

_____

_____

IV. CHARLES' LAW: THE TEMPERATURE-VOLUME LAW

Data to be recorded (all volumes are in mL):

Volume **A** (that of water and equipment, which includes empty balloon) _____

Volume **B** (that of water, equipment, and air sample at 0°C) _____

Volume **C** (volume of air sample at 0°C) _____

Volume **D** (that of water, equipment, and air sample at _____°C) _____

Volume **E** (**observed** volume of air sample at the higher temperature) _____

Volume **F** (**calculated** volume of air sample at the higher temperature) _____

On the absolute temperature scale, the higher temperature is equal to the temperature in Celsius recorded above plus 273. So it is _____ K.

Volume **F** = (volume **C**)(_____K/273K) = _____ mL

Comparison of observed volume **E** vs. calculated volume **F**:

1. Percentagewise, how much larger or smaller is **E** than **F**?

2. If there is a discrepancy, how might it be explained?

V. HENRY'S LAW: THE PRESSURE-SOLUBILITY LAW

1. a. What happened to the carbonated water when the pressure in the suction flask decreased?

   _____

   b. Explain your observations in *a*.

   _____

   _____

2. a. What happened to the carbonated water when the vacuum tubing was disconnected from the aspirator?

   _____

   b. Explain your observations in *a*.

   _____

   _____

VI. GAY-LUSSAC'S LAW: THE PRESSURE-TEMPERATURE LAW AND ABSOLUTE ZERO (Instructor's Optional Demonstration)

1 Data (Remember to include units with each quantity):

Temperature	Pressure
_____	_____
_____	_____
_____	_____

2. a. Plot the data on graph, Figure 9-5. What is the indicated value of absolute zero?

_____

b. How does your experimental value compare with the accepted value of absolute zero?

_____

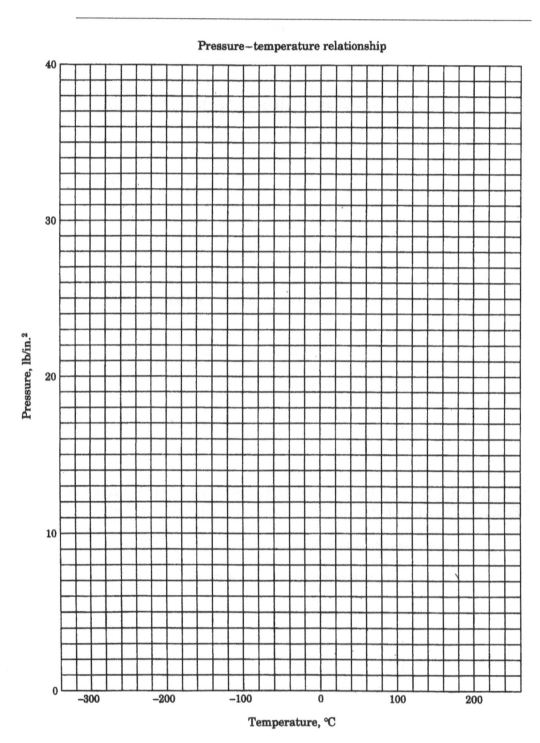

Pressure–temperature relationship

Figure 9-5 The relationship between pressure and temperature of a gas. Absolute zero is the temperature at which the pressure is zero.

Related Questions for Experiment 9

1. Use the *CRC Handbook of Chemistry and Physics* to find the solubility of carbon dioxide in water. It will be in the section "Physical Constants of Inorganic Compounds" of the 75th (1994–95) edition or earlier editions. This information can also be found in *The Merck Index*.

    a. Solubility in cold water (0°C) _____ cm^3 CO_2 per 100 cm^3 H_2O

    b. Solubility in hot water (20°C) _____ cm^3 CO_2 per 100 cm^3 H_2O

    c. Solubility in cold water (0°C) _____ g CO_2 per 100 cm^3 H_2O

    d. Solubility in hot water (60°C) _____ g CO_2 per 100 cm^3 H_2O

    e. This difference seen in solubility at different temperatures is typical of gases. How does increasing temperature affect the solubility of a gas in water?

    _____

    _____

2. "Thermal pollution" is the term used to describe hot water added to a lake or river by an electrical generating plant or other sources. Why is increased water temperature an environmental concern?

    _____

    _____

3. When deep-sea divers ascend too quickly, they suffer severe blood circulation impairment due to what is known as "the bends." Which portion of this experiment illustrates the cause of the bends?

4. What can you assume about the relative solubility of helium and nitrogen since supplying a mixture of helium and oxygen gases to deep-sea divers rather than compressed air (nitrogen and oxygen) prevents the bends?

5. An old method of treating the bends was to pack the diver in ice. Explain why lowering the body temperature served as an emergency treatment.

    _____

    _____

6. In a tragic 1989 accident, a cargo door of an airplane fell off as the plane flew at 22,000 feet over the South Pacific. As the door fell away, it ripped out a large hole in the passenger cabin. Nine passengers were swept out of the opening and many more were hurled toward the opening but managed to stay inside the plane. Use the gas laws to explain why the passengers were affected in this way.

_____

_____

7. Explain why aerosol cans should be stored in a cool place.

_____

_____

Experiment 10

Acids, Bases, Buffers, and Antacids

References: *General, Organic, and Biological Chemistry: An Integrated Approach, 4th ed., Chapter 7.*

Purpose: To study the behavior of acids, bases, buffers, and antacids. To determine the effectiveness of a commercially available antacid.

Introduction

ACIDS AND BASES

Acids and bases can be defined in several different ways. Generally, the most useful is the Brønsted-Lowry definition which states that an **acid** is a substance capable of donating a proton. A substance capable of accepting a proton is a **base**.

When added to water, all acids can donate a proton (H^+) to water; the products of the reaction will be a hydronium ion (H_3O^+) and a negative ion. Because ions are formed, all acids are electrolytes. **Strong acids** are almost completely ionized in water and are therefore strong electrolytes. Strong acids easily donate protons to acceptors. Nitric acid, mentioned earlier, is a strong acid. Only a small number of common acids are strong acids: hydrochloric acid (HCl), sulfuric acid (H_2SO_4), and nitric acid (HNO_3).

$$HNO_3 \ (aq) \ + \ H_2O \ (l) \ \rightleftharpoons \ H_3O^+ \ (aq) \ + \ NO_3^- \ (aq)$$
$$< 1\% \hspace{5cm} > 99\%$$

Weak acids are only slightly ionized in water and are weak electrolytes. Weak acids donate protons, but with limited success. Acetic acid is a weak acid. Most common acids are weak acids.

$$CH_3COOH \ (aq) \ + \ H_2O \ (l) \ \rightleftharpoons \ H_3O^+ \ (aq) \ + \ CH_3COO^- \ (aq)$$
$$99\% \hspace{5cm} 1\%$$

Bases, too, are either strong or weak. **Strong bases** are nearly totally ionized in water and hence are strong electrolytes; **weak bases** are weak electrolytes because they only partially ionize in water. Strong bases are better proton acceptors than weak bases. Sodium hydroxide ($NaOH$) and potassium hydroxide (KOH) are common strong bases; both ionize in water to release hydroxide ion (OH^-). The hydroxide ion is an excellent proton (H^+) acceptor.

$$OH^- \ (aq) \ + \ H_3O^+ \ (aq) \ \longrightarrow \ 2 \ H_2O \ (l)$$

One of the most common weak bases is ammonia (NH_3). It contains no hydroxide ion but to a very limited extent reacts with water, forming an ammonium ion (NH_4^+) and a hydroxide ion.

$$NH_3 \ (aq) \ + \ H_2O \ (l) \ \rightleftharpoons \ NH_4^+ \ (aq) \ + \ OH^- \ (aq)$$
$$99\% \hspace{5cm} 1\%$$

NEUTRALIZING REACTIONS

In Experiment 7, you investigated one of the common reactions of acids—the reaction with a reactive metal to produce hydrogen gas. Another property of acids is that they react with and neutralize bases, producing a neutral solution containing a salt and water. Table salt (NaCl) is the most common salt but many other substances, such as KCl, are also salts. A **salt** is an ionic compound and is one of the products of an **acid-base neutralization reaction**. Because salts are ionic, they are electrolytes. All electrolytes are either an acid, a base, or a salt.

$$acid \ + \ base \ \longrightarrow \ salt \ + \ water$$

A solution of potassium hydroxide is basic. As hydrochloric acid is added to the solution, the acid consumes the base, converting it to a salt (KCl) and water. As the base is destroyed, the solution becomes neutral. If additional acid is added, the solution becomes acidic.

$$HCl \ (aq) \ + \ KOH \ (aq) \ \longrightarrow \ KCl \ (aq) \ + \ H_2O \ (l)$$

Acids are neutralized by bases in the same manner. A solution of hydrochloric acid is, of course, acidic. If potassium hydroxide is added to the solution, the acid is consumed and the solution turns neutral. Additional potassium hydroxide will make the solution basic.

THE pH SCALE

The pH scale is a convenient way of indicating the acidity or basicity of aqueous solutions. Pure water and all neutral solutions contain equal concentrations of hydronium ions and hydroxide ions, $[H_3O^+] = [OH^-]$, and the pH is 7. An acidic solution is one in which the concentration of hydronium ions is greater than the concentration of hydroxide ions, $[H_3O^+] > [OH^-]$ and the pH is less than 7. When $[OH^-] > [H_3O^+]$, the solution is basic (alkaline) and the pH is greater than 7.

BUFFERS

Buffers are solutions that resist a change in pH. They protect against drastic changes in pH when acids or bases are added to the solution. The body's blood and extracellular fluids contain buffers. The buffering capacity of a system may be overloaded if huge amounts of acid or base are added; in the blood, this can lead to acidosis or alkalosis.

The buffer used in this experiment is a mixture of potassium dihydrogenphosphate (KH_2PO_4) and potassium monohydrogenphosphate (K_2HPO_4). A solution of this mixture contains the following equilibrium:

$$H_2PO_4^- \ (aq) \ + \ H_2O \ (l) \ \rightleftharpoons \ HPO_4^{2-} \ (aq) \ + \ H_3O^+ \ (aq)$$

If acid is added to the buffer solution, the stress of added hydronium ion causes the equilibrium to shift to the left, relieving the stress. Most of the additional hydronium ion is changed into water and the pH stays relatively constant.

Bases decrease the concentration of hydronium ion in a solution. If base is added to a buffer solution, the stress of reduced hydronium ion concentration causes the equilibrium to shift to the right. Water is converted to hydronium ion and the pH does not significantly change.

A. The pH Meter

The pH meter provides the most accurate means of measuring the pH of a solution. pH meters are calibrated using standard buffer solutions; then the electrodes are dipped into the solution to be tested and a pH reading is made from the meter. Many different types of pH meters are available; your instructor will describe the use of your pH meter.

B. Indicators

Another simpler, but less accurate, method of measuring pH uses chemical dyes called indicators. **Indicators** change color at certain pH levels.

One commonly used indicator is phenolphthalein, which is colorless in acid and red in base. Another is bromphenol blue (yellow in acid, blue in base). It is used in estimating the buffering capacity of antacids because it reaches its in-between color of green at about pH 3, that of stomach fluids.

Litmus is a naturally occurring dye that is either red or blue. Strips of **litmus paper** are commonly used to indicate whether a solution is acidic (pH < 7), basic (pH > 7), or neutral (pH = 7). Blue litmus paper turns pink when placed in an acidic solution, and red litmus paper turns blue in basic solution (remember this as $B=B$, if red paper turns **b**lue, the solution is **b**asic). If neither paper changes color, the solution is neutral.

Universal indicator, a mixture of dyes, is not used in this experiment but will be used in subsequent experiments. Several different mixtures of universal indicator are manufactured; the most common one displays the following colors when added to a solution.

pH	Color	pH	Color
4	Rose	8	Blue-green
5	Gold	9	Blue
6	Yellow	10	Violet
7	Green		

pH paper contains a mixture of indicators which turn a range of colors when wetted with a drop of solution. Each type of pH paper must be used in conjunction with the supplied table of colors to determine the pH of the tested solution. The most frequently used pH paper covers a wide pH range (1–12) but is accurate within only one pH unit. Other types of pH paper measure narrower pH ranges but will supply a greater degree of accuracy.

HEARTBURN

The digestion of food is a mechanical and chemical process that breaks complex organic compounds into simple soluble substances absorbable by tissues. Digestion begins in the mouth and continues in the stomach and the small intestines. The digestive reactions of the stomach are catalyzed by enzymes found in the gastric juice secreted by the stomach. These enzymes are initially in an inactive form and are activated by hydrochloric acid, separately secreted into the stomach. A secreted layer of mucus protects the stomach walls from attack by the gastric juices. The mixture of food and gastric juice, called chyme, is normally prevented from backing up into the esophagus, a nine-inch-long passageway from the throat to the stomach. The esophagus is not protected by mucus and it can be painfully irritated by refluxing chyme if the esophageal muscle does not completely close. This failure can result from overeating, stress, pregnancy, or lying down soon after a meal. Fatty foods, alcohol, and chocolate adversely affect the muscle's ability to completely close; spicy foods can directly irritate the esophagus. The irritation of the esophagus by chyme results in a stinging feeling behind the breastbone. This pain is sometimes referred to as **heartburn**, **acid indigestion**, or **sour stomach**. Heartburn pain is likely when chyme of pH below 3 backs up into the esophagus.

Antacids are bases that neutralize some of the hydrochloric acid in the stomach. To be effective, an antacid should quickly raise the pH above 3 and hold it there for a period of time. Some brands of antacids will neutralize greater quantities of acid than others, some act more quickly, some are longer lasting, and some have fewer side effects. None are ideal. Most antacids include one or more of the following four bases:

Sodium bicarbonate, ordinary baking soda, is fast-acting and inexpensive. Although a good choice for occasional use, sodium bicarbonate is the least desirable for frequent use. Regular use may throw off the body's acid-base balance and encourage urinary infections and kidney disorders. People on low-sodium diets need to avoid antacids containing sodium bicarbonate. Sodium bicarbonate neutralizes stomach acid in the following manner:

$$NaHCO_3 + HCl \longrightarrow NaCl + H_2O + CO_2 (g)$$

The buildup of carbon dioxide gas pressure may eventually lead to a "burp of relief."

Calcium carbonate (chalk) is also a potent, fast-acting, inexpensive antacid that is good for occasional use. Some brands of antacids are also promoted as calcium supplements. The maximum recommended daily intake of calcium is 3200 mg; prolonged excessive calcium intake may cause urinary tract damage, kidney stones, reduced iron levels, drug interactions, constipation, and confusion. Hydrochloric acid is neutralized by the process:

$$CaCO_3 + 2 HCl \longrightarrow CaCl_2 + H_2O + CO_2 (g)$$

Magnesium hydroxide, also known as milk of magnesia, is a good antacid for most people. Magnesium hydroxide has laxative effects and is often combined with calcium or aluminum compounds which are constipating. People with kidney problems must avoid long-term use of magnesium antacids—very serious problems with respiration, heartbeat, and blood pressure are possible. The neutralization reaction is:

$$Mg(OH)_2 + 2 HCl \longrightarrow MgCl_2 + 2 H_2O$$

Aluminum hydroxide is slow-acting, but it provides long-lasting antacid action. It is frequently combined with magnesium hydroxide to counter aluminum hydroxide's constipating properties. People with kidney disease should consult a physician before taking antacids containing aluminum.

$$Al(OH)_3 + 3 HCl \longrightarrow AlCl_3 + 3 H_2O$$

Most hydroxides are damaging to tissues, but those of magnesium and aluminum are less so because of their limited solubilities.

In addition to these ingredients, almost all antacids contain sweeteners, typically saccharin or sugar. Some antacids contain aspirin—the aspirin would help a headache but may cause additional stomach irritation. Other brands contain caffeine, which may stimulate additional secretion of stomach acid. Simethicone is sometimes included to reduce stomach gas but there is little evidence that this is an effective treatment. Gaviscon® antacid forms a foamy floating barrier in the stomach to prevent backup into the esophagus. Some stomach products, such as Pepto-Bismol®, are not antacids at all, and promise "soothing relief" through a "protective coating action" of bismuth subsalicylate. Some over-the-counter products, such as Pepcid AC®, Tagamet HG®, and Zantac 75®, are acid blockers rather than antacids. They reduce the secretion of stomach acid by inhibition of the enzyme that is responsible for pumping H^+ into the stomach.

Over time, excess stomach acid can cause more than heartburn. Acid can damage the stomach lining and cause ulcers or inflammation (gastritis). Because of the side-effects of antacids and the possibility that the symptoms could be due to stomach cancer or ulcers, it is wise to see a physician before starting frequent or long-term antacid use.

EFFECTIVENESS OF ANTACIDS

Stomach acid is approximately 0.10 M hydrochloric acid. In attempting to evaluate the effectiveness of antacids, the simple and direct approach of adding 0.10 M HCl to an antacid tablet will not work because many antacids are buffered. The presence of a buffer makes it difficult to determine when all the antacid has been consumed. This problem can be avoided with an indirect analysis. You will add more hydrochloric acid than is necessary to react with the antacid. Only part of the HCl will be neutralized by the antacid.

$$HCl \ + \ antacid \ \longrightarrow \ salt \ + \ H_2O$$

The quantity of HCl not neutralized by antacid can be determined by neutralizing the excess HCl with sodium hydroxide (NaOH).

$$HCl \ + \ NaOH \ \longrightarrow \ NaCl \ + \ H_2O$$

The process of determining the amount of a solution required to react with a given amount of a sample is called a **titration**. In this experiment, a **back titration** is used to determine the amount of HCl that was not neutralized by the antacid. Bromphenol blue is an indicator that is yellow below pH 3, blue at pH values near 5, and changes to green in between those two values. It will be used in this experiment to determine when the HCl solution has been neutralized to a pH above 3 similar to the action of an antacid in the stomach.

Consider the following data:

Mass of antacid	5.00 g
Recommended dose	2 tablets
Volume of HCl added to 1 tablet	50.0 mL
Volume of NaOH used in titration	10.00 mL

A total of 50 mL of 0.10 M HCl was added to the antacid, some of it reacted with the antacid and some was left over, unreacted. The amount of excess, unreacted HCl was determined by titrating with 0.10 M NaOH solution. Since 1 mole of NaOH neutralizes 1 mole of HCl and since the 0.10 M NaOH has the same molar concentration as the 0.10 M HCl, every milliliter of NaOH will neutralize 1 milliliter of HCl.

$$\left(\begin{array}{c} \text{volume of HCl} \\ \text{neutralized by NaOH} \end{array} \right) = \left(\begin{array}{c} \text{volume of NaOH} \\ \text{used in titration} \end{array} \right)$$

If 50 mL of HCl were added to the antacid and 10 mL were neutralized by the NaOH, how much HCl was neutralized by the antacid?

$$\begin{pmatrix} \text{volume of HCl neutralized} \\ \text{by 1 tablet of antacid} \end{pmatrix} = \begin{pmatrix} \text{volume of HCl} \\ \text{added to sample} \end{pmatrix} - \begin{pmatrix} \text{volume of HCl} \\ \text{neutralized by NaOH} \end{pmatrix}$$

volume of HCl neutralized by 1 tablet of antacid = 50 mL – 10 mL = 40 mL

The recommended dosage of antacids varies by brand; the dose for the tested brand is two tablets. If 40 mL of HCl were neutralized by 1 antacid tablet, what volume of HCl was neutralized by 1 dose of antacid? This is sometimes referred to as the **Acid Neutralizing Capacity (ANC)** per dose.

$$ANC = \begin{pmatrix} \text{volume of HCl} \\ \text{neutralized by} \\ \text{1 dose of antacid} \end{pmatrix} = \begin{pmatrix} \text{volume of HCl} \\ \text{neutralized by} \\ \text{1 tablet of antacid} \end{pmatrix} \times \begin{pmatrix} \text{number of} \\ \text{tablets} \\ \text{per dose} \end{pmatrix}$$

$$ANC = \frac{\text{volume of HCl neutralized}}{\text{by 1 dose of antacid}} = \frac{\text{40 mL HCl neutralized}}{\text{1 tablet}} \times \frac{\text{2 tablets}}{\text{1 dose}}$$

$$ANC = \frac{\text{volume of HCl neutralized}}{\text{by 1 dose of antacid}} = \frac{\text{80 mL HCl neutralized}}{\text{1 tablet}}$$

The ANC is a method of comparing the effectiveness of various brands of antacids. This is the method that will be used in this experiment.

Procedure

Disposal: The products of all procedures may be poured down the drain *EXCEPT* for any left over 0.1 M HCl (e.g., that in Part II.B), which should go in a "Waste Acids" container.

I. MEASUREMENT OF pH

A. Acidity and Basicity of Substances

1. Label nine wells in a white porcelain well plate from 1 to 9.

2. Put about 0.5 mL of each of the below aqueous solutions into the identically numbered well:

> No. 1: 0.9% sodium chloride (NaCl; isotonic saline)
> No. 2: 0.1 M trisodium phosphate (Na$_3$PO$_4$)
> No. 3: 10% ethanol (CH$_3$CH$_2$OH)
> No. 4: 0.1 M ammonia (NH$_3$)
> No. 5: 0.1 M sodium hydroxide (NaOH)
> No. 6: 5% acetic acid (CH$_3$COOH; vinegar)
> No. 7: 0.1 M sodium hydrogen carbonate (NaHCO$_3$)
> No. 8: 0.1 M sodium carbonate (Na$_2$CO$_3$)
> No. 9: 0.1 M hydrochloric acid (HCl)

3. Test the pH of the solutions by dipping a stirring rod in and then touching the rod on a short strip of wide-range pH test paper (pH range 1-12). Wash the rod after each test. Record the data in the Report section.

B. pH of Some Foods

1. Label five other (or washed) wells in the well plate from 10 to 14.

2. Put about 0.5 mL of each of the below beverage solutions into the identically numbered well:

   No. 10: Orange juice
   No. 11: Lemon juice
   No. 12: Tomato juice
   No. 13: Tea
   No. 14: Soda water

3. Test the pH of the solutions as done above.

II. ACID-BASE NEUTRALIZATION

1. Pour about 30 mL of 0.1 M sodium hydroxide (NaOH) solution into a labeled 100-mL beaker.

2. Pour about 30 mL of 0.1 M hydrochloric acid (HCl) solution into a labeled 100-mL beaker.

3. Transfer 1 mL of the 0.1 M HCl to a 16 × 150-mm test tube.

4. Add one drop of phenolphthalein solution to the test tube. Record the color of the solution.

5. While stirring the solution, add 0.1 M NaOH from an eye dropper one drop at a time. Count and record the drops added until the solution turns just perceptively pink.

6. Add one or two additional drops of NaOH and record the color.

7. Using a clean eye dropper, add 0.1 M HCl to the test tube, one drop at a time, until the solution again becomes pale pink. The solution is neutral at this point.

8. Save the rest of the NaOH and HCl solutions for use in Part III.

III. BUFFERS

A. The Buffering Ability of Water

1. Review Fundamental Laboratory Operation, Part 3c, pages 13-15, for the preparation and use of burets.

2. Clean two 25- or 50-mL burets with a buret brush, rinse well with tap water, allowing the water to drain through the tip. Rinse with a few milliliters of deionized water.

3. Place both burets in a buret clamp. Label one buret "NaOH" and the other "HCl."

4. Using the 0.1 M NaOH solution from Part II, pour a few milliliters of NaOH solution into the first buret. Discard the solution through the tip since the solution is diluted with rinse water.

5. Pour about 20 mL of 0.1 M NaOH into the buret. Allow a small amount of NaOH solution to drain through the tip until there is a solid column of solution extending down to the tip, completely free of air bubbles.

6. Adjust the level of NaOH solution to the 15-mL mark on the buret. Remember to read the bottom of the meniscus.

7. Pour 25 mL of distilled water or boiled deionized water into a 150-mL beaker. Use a pH meter to record the pH.

8. Use the buret to add 0.5 mL of NaOH to the beaker of deionized water. Stir the mixture. Record the pH.

9. Add an additional 0.5 mL of NaOH solution. Stir and record the pH.

10. Discard the solution in the beaker as directed by the instructor; wash and dry it thoroughly. Thoroughly rinse the electrodes of the pH meter.

11. Using the second buret, repeat steps 4 through 10 using 0.1 M HCl in place of NaOH.

12. Save the 0.1 M solutions of NaOH and HCl in the burets for Part B.

B. The Buffering Ability of a Phosphate Buffer

1. Pour 25 mL of a KH_2PO_4/K_2HPO_4 buffer solution into a 150-mL beaker. Record the pH using a pH meter.

2. Use the buret to add 0.1 M NaOH in 0.5 mL quantities until a volume of 5 mL has been added. Stir after each addition and record the pH.

3. Discard the solution in the beaker as directed by the instructor; wash and dry it thoroughly. Thoroughly rinse the electrodes of the pH meter.

4. Pour 25 mL of a KH_2PO_4/K_2HPO_4 buffer solution into a 150-mL beaker. Record the pH using a pH meter.

5. Use the buret to add 0.1 M HCl in 0.5 mL quantities until a volume of 5 mL has been added. Stir after each addition and record the pH.

6. Save the 0.1 M NaOH buret for use in Part IV. Discard the solution from the 0.1 M HCl buret in a "Waste Acids" container. If requested, rinse the burets thoroughly with tap water, allowing the water to drain through the tip. Rinse with a few mL of deionized water.

7. Thoroughly rinse the electrodes of the pH meter.

IV. ANTACIDS

In Part IV, better results will be obtained if the HCl and NaOH solutions have more precisely known concentrations, e.g., 0.10 M, but it can be done with 0.1 M.

1. Determine the mass of a sheet of weighing paper and record the mass on the Report. If the balance has an automatic taring feature, you may wish to use this feature and enter TARE or 0.00 g for the mass of the weighing paper.

2. Place an antacid tablet on the weighing paper and determine its mass. <u>Record the brand name and the mass of the antacid.</u> White tablets are easier to titrate than colored tablets.

3. Use a mortar and pestle to crush the antacid tablet. Transfer the crushed tablet to a 250-mL Erlenmeyer flask. Rinse the powder remaining in the mortar into the flask with 5–10 mL of deionized water.

4. From a 50-mL graduated cylinder, add exactly 50 mL of 0.1 M HCl. Swirl the flask to dissolve as much of the crushed tablet as possible.

5. Add 5–10 drops of bromphenol blue indicator to the flask. If the solution is yellow, record the amount of HCl added and proceed to Step 6. If the solution is blue, add 25-mL quantities of HCl (swirl the flask each time!) until a yellow color is obtained; record the total volume of HCl added.

6. Using the 0.1 M NaOH buret employed in Part III, adjust the level of the NaOH solution to the zero mark with 0.1 M NaOH. Record the initial reading of NaOH in the buret. Remember to read the bottom of the meniscus.

7. Use a sheet of paper under the flask to provide a white background. Titrate the yellow antacid mixture to a green endpoint. Swirl the flask as NaOH is added. The green color should persist for 30 seconds. If a colored antacid tablet was used, the endpoint may not be green; the endpoint for colored antacids is any change in color. Record the final reading of NaOH in the buret. Calculate the volume of NaOH used in the titration.

8. Calculate the volume of HCl neutralized by the antacid.

9. Clean-up: it is important that the NaOH solution be thoroughly rinsed from the buret. Dispose of all solutions in the "Waste Salts" container.

Acids, Bases, Buffers, and Antacids

Prelab Questions for Experiment 10

_____ 1. Which of the following will provide the most accurate means of measuring pH?
   a. bromphenol blue
   b. litmus paper
   c. pH paper
   d. a pH meter

_____ 2. When a drop of a solution is placed on red litmus paper, the paper turns blue. The solution is
   a. acidic.
   b. basic.
   c. neutral.

_____ 3. A solution becomes yellow after the addition of one drop of bromphenol blue. The solution is
   a. acidic.
   b. basic.
   c. neutral.

_____ 4. After adding 5 mL of acid to a solution having a pH of 9.6, the pH changed to 9.5. The solution is
   a. acidic.
   b. a buffer.
   c. a nonelectrolyte.
   d. a weak electrolyte.

_____ 5. Which statement about a pH 5 solution is correct?
   a. The solution is neutral.
   b. The solution is a buffer.
   c. The hydroxide ion concentration is greater than the hydronium ion concentration.
   d. The hydronium ion concentration is greater than the hydroxide ion concentration.

6. a. When using bromphenol blue to titrate an acid, what color will the indicator be prior to titration with NaOH solution? _____

   b. What color will it be on reaching the normal pH of the stomach? _____

7. Given the following data, fill in the blank:

Brand of antacid	Brand X
Recommended dose	2 tablets
Mass of 1 antacid and weighing paper	2.50 g
Mass of weighing paper	0.20 g
Mass of 1 antacid tablet	
Volume of HCl added to sample	75.0 mL
Final reading of buret (NaOH)	14.50 mL
Initial reading of buret (NaOH)	0.00 mL
Volume of NaOH used in titration	
Volume of HCl neutralized by 1 tablet of antacid	
ANC (volume of HCl neutralized by 1 dose of antacid)	

Calculations:

Report for Experiment 10

I. MEASUREMENT OF pH

A. Acidity and Basicity of Substances

1. Enter data here:

Solution	pH	Solution	pH
1. 0.9% NaCl		6. 5% CH_3COOH	
2. 0.1 M Na_3PO_4		7. 0.1 M $NaHCO_3$	
3. 10% CH_3CH_2OH		8. 0.1 M Na_2CO_3	
4. 0.1 M NH_3		9. 0.1 M HCl	
5. 0.1 M NaOH			

2. Write the formulas of the:

Weak acids _____

Weak bases _____

B. pH of Some Foods

1. Enter data here:

Solution	pH	Solution	pH
10. Orange juice		13. Tea	
11. Lemon juice		14. Soda water	
12. Tomato juice			

2. Which beverage was least acidic? _____

3. Which beverage was most acidic? _____

4. Which solution in section A could be considered a beverage if it were not denatured (<u>as it is</u>)? _____

II. ACID-BASE NEUTRALIZATION

    1. Data:

Color of phenolphthalein indicator in HCl solution (Step 4)   _____

Number of drops of NaOH to turn solution pale pink (Step 5)   _____

Color of solution with 1 or 2 additional drops of NaOH (Step 6)   _____

    2. Is the color change reversible (can you change the color back and forth)?

_____

    3. Write a complete, balanced equation for the neutralization of hydrochloric acid by sodium hydroxide.

_____

III. BUFFERS

A. The Buffering Ability of Water

    1. Data:

mL of NaOH	pH	mL of HCl	pH
0.0		0.0	
0.5		0.5	
1.0		1.0	

    2. A buffer is a substance that resists changes in pH when a small amount of acid or base is added. Is water an effective buffer?

_____

B. The Buffering Ability of a Phosphate Buffer

1. Data:

mL of NaOH	pH		mL of HCl	pH
0.0			0.0	
0.5			0.5	
1.0			1.0	
1.5			1.5	
2.0			2.0	
2.5			2.5	
3.0			3.0	
3.5			3.5	
4.0			4.0	
4.5			4.5	
5.0			5.0	

2. Is the KH_2PO_4/K_2HPO_4 mixture a more effective buffer than water?

_____

3. Was there evidence that the buffering capacity was used up by addition of large amounts of acid or base? Explain.

_____

_____

4. Label the conjugate acid-base pairs in the acid-base equilibrium of the buffer system:

$$H_2PO_4^- \; (aq) \;\; + \;\; H_2O \; (l) \;\; \rightleftharpoons \;\; HPO_4^{2-} \; (aq) \; + \; H_3O^+ \; (aq)$$

IV. ANTACIDS

1. Data:

Remember to include units with each quantity.

Brand of antacid	
Recommended dose	
Cost of 1 tablet	
Cost of 1 dose	
Mass of 1 antacid tablet and weighing paper	
Mass of weighing paper	
Mass of 1 antacid tablet	
Volume of HCl added to sample	
Final reading of buret (NaOH)	
Initial reading of buret (NaOH)	
Volume of NaOH used in titration	
Volume of HCl neutralized by 1 tablet of antacid	
ANC (volume of HCl neutralized by 1 dose of antacid)	

Calculations:

2. Ingredients:

Brand Name of antacid studied. _____

Ingredients of this antacid.

_____

_____

_____

3. In addition to acid neutralizing capacity, what other considerations are important in selecting an antacid?

_____

_____

Acids, Bases, Buffers, and Antacids

Related Questions for Experiment 10

1. Acidosis is a condition in which the blood pH drops below normal. Acidosis can be caused by metabolic or respiratory conditions. Explain the role of carbon dioxide in producing respiratory acidosis. Hint: Recall the pH you found for soda water and consult your text. (Chapter 7 in *General, Organic, and Biological Chemistry: An Integrated Approach, 4th ed.*)

_____

_____

2. A person who is hyperventilating suffers from alkalosis, a condition in which the blood serum pH is too high. Explain why breathing into a paper bag is a treatment for hyperventilation.

_____

_____

3. An upset stomach can be caused by excess stomach acid. Why can baking soda ($NaHCO_3$) settle an upset stomach?

_____

_____

4. a. Which substance, $H_2PO_4^-$ or HPO_4^{2-}, would appear in the urine in alkalosis?

   _____

   b. In acidosis?

   _____

   c. Explain.

   _____

   _____

   _____

5. A student measured the pH of coffee and obtained the following results:
   caffeinated coffee   pH 5.6
   decaffeinated coffee   pH 5.0

   a. On the basis of this data, would you classify caffeine as an acid or base? _____

   b. Explain.

   _____

   _____

6. Bacteria in the mouth convert sugar into acids that are capable of dissolving tooth enamel. Repeated drinking of carbonated beverages, even those that do not contain sugar, can also contribute to loss of enamel. Explain why carbonated beverages can act in this way and cause dental caries (tooth decay).

_____

_____

7. Classify each of the following acids as strong or weak acids (S or W):

_____ HNO_3

_____ H_3PO_4

_____ H_2CO_3

_____ citric acid

_____ sulfuric acid

_____ ascorbic acid

8. Complete the following table to illustrate the difference between acid strength and acid concentration:

	Strong or Weak Acid? (S or W)	Concentrated or Dilute Acid? (C or D)
12 M HCl (hydrochloric acid)		
0.1 M HCl		
17 M CH_3COOH (glacial acetic acid)		
0.1 M CH_3COOH		

Experiment **11**

Organic Oxygen Compounds

Reference: *General, Organic, and Biological Chemistry: An Integrated Approach, 4th ed., Chapters 4, 8, and 9.*

Purpose: You will investigate physical and chemical properties of organic compounds containing oxygen. You will perform classification tests for aldehydes and acetone that are frequently used in medicine. Two esters will be synthesized. During the course of the experiment, you will identify the compounds you encounter as being components of commercially available products.

Introduction

The physical and chemical properties of the organic oxygen compounds are discussed in detail in your text. The simple test tube experiments you will perform illustrate some of the reactions most relevant to body chemistry. For example, the oxidation of ethanol that you will carry out in a test tube is similar in many respects to its oxidation in the liver. We shall later see that the oxidation of an aldehyde with a mild oxidizing agent is the technique used to detect the presence of sugar in the urine. In later experiments we will use a modification of Legal's test for identifying acetone in urine—the tell-tale symptom of diabetes.

ALCOHOLS

Alcohols are classified as primary, secondary, or tertiary according to the placement of the hydroxyl group on the molecule. Primary (1°) alcohols have the hydroxyl group attached to a carbon atom that is bonded to one other carbon atom.

$$RCH_2OH \qquad\qquad R_2CHOH \qquad\qquad R_3COH$$
$$1° \text{ alcohol} \qquad\qquad 2° \text{ alcohol} \qquad\qquad 3° \text{ alcohol}$$

A secondary alcohol (2°) has the hydroxyl group bonded to a carbon atom that is bonded to two other carbon atoms. A tertiary (3°) alcohol has the hydroxyl group attached to a carbon atom that is bonded to three other carbon atoms.

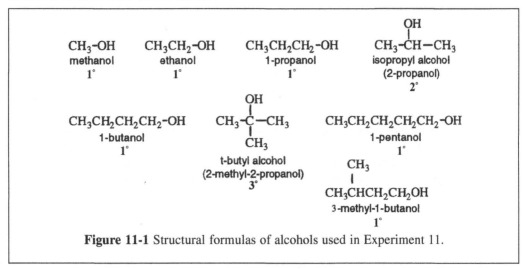

Figure 11-1 Structural formulas of alcohols used in Experiment 11.

The hydroxyl group in an alcohol molecule forms a polar region and permits strong hydrogen bonding between molecules. This explains why alcohol boiling points are much higher than those of their parent hydrocarbons, e. g., methanol boils at 65 °C vs. −164 °C for methane. Hydrogen bonding with water molecules explains the water solubility of the alcohols whose molecules contain no more than four carbon atoms per hydroxyl group. With more carbon atoms per hydroxyl, the alkyl group becomes predominant and water solubility rapidly decreases.

Phenols (aromatic alcohols) have a hydrogen atom on a benzene ring replaced by a hydroxyl group and are generally water-insoluble. However, because they are more acidic than ordinary alcohols (but less so than carboxylic acids), they dissolve in dilute base because of the formation of a salt.

$$OH O^- \quad Na^+$$

$$\bigcirc + NaOH \longrightarrow \bigcirc + H_2O \qquad \text{(Eq. 1)}$$

water insoluble water soluble

Phenols are powerful skin irritants; the troublesome oil on the leaves of plants like poison ivy and oak is urushiol, which has two phenolic hydroxyl groups and a long-chain alkyl group on its benzene ring.

In this experiment you will synthesize carboxylic acids (RCOOH) and ketones (R_2CO) by oxidizing a primary or secondary alcohol.

Primary alcohols can be oxidized to form aldehydes as shown in Equation 2. With most oxidizing agents, the intermediate aldehyde is quickly further oxidized to the carboxylic acid as in Equation 3. In organic chemistry, an **oxidation** is the gain of an oxygen atom or the loss of two hydrogen atoms. The primary alcohol of Equation 2 was oxidized because it lost two hydrogen atoms. The aldehyde of Equation 3 gained an oxygen atom in its oxidation.

$$
\begin{array}{c}
O\!-\!H \\
| \\
R\!-\!C\!-\!H \\
| \\
H
\end{array}
+ [O] \longrightarrow
\begin{array}{c}
O \\
\| \\
R\!-\!C\!-\!H
\end{array}
+ H\!-\!O\!-\!H \qquad \text{(Eq. 2)}
$$

1° alcohol $\quad$ oxidizing agent aldehyde water

$$
\begin{array}{c}
O \\
\| \\
R\!-\!C\!-\!H
\end{array}
+ [O] \longrightarrow
\begin{array}{c}
O \\
\| \\
R\!-\!C\!-\!O\!-\!H
\end{array}
\qquad \text{(Eq. 3)}
$$

aldehyde $\quad$ oxidizing agent carboxylic acid

The alcohol and aldehyde in Equations 2 and 3 are oxidized by an **oxidizing agent**, symbolized by [O]. In this experiment, the oxidizing agent is sodium dichromate ($Na_2Cr_2O_7$) in sulfuric acid (H_2SO_4). Sodium dichromate is a very strong oxidizing agent and the aldehyde formed in Equation 2 cannot be isolated; the intermediate aldehyde is immediately further oxidized to a carboxylic acid. The overall reaction with sodium dichromate and sulfuric acid is shown in Equation 4.

$$3\ RCH_2OH + 2\ Na_2Cr_2O_7 + 10\ H_2SO_4 \longrightarrow 3\ RCO_2H + 11\ H_2O + 4\ NaHSO_4 + 2\ Cr_2(SO_4)_3 \quad \text{(Eq. 4)}$$
1°alcohol   (orange)   (carboxylic acid)   (green)

When it reacts, the bright orange color of chromium(VI) in sodium dichromate is changed to the green chromium(III) ion, Cr^{3+}. Hence if a reaction takes place, there will be a sharp color change in the solution. Equation 4 is complex; we can simplify it by omitting some of the inorganic reactants and products and leaving it unbalanced (Equation 5).

$$RCH_2OH + Na_2Cr_2O_7 \longrightarrow RCO_2H + Cr_2(SO_4)_3 \quad \text{(Eq. 5)}$$
1° alcohol   (orange)   carboxylic acid   (green)

Secondary alcohols react similarly to form ketones (Equations 6 and 7).

$$\text{(Eq. 6)}$$

2° alcohol   oxidizing agent   ketone   water

$$R_2CHOH + Na_2Cr_2O_7 \longrightarrow RCOR + Cr_2(SO_4)_3 \quad \text{(Eq. 7)}$$
2° alcohol   (orange)   ketone   (green)

Tertiary alcohols do not react and leave a clear orange solution of unreacted sodium dichromate (Equations 8 and 9).

$$\text{(Eq. 8)}$$

3° alcohol   oxidizing agent

$$R_3COH + Na_2Cr_2O_7 \longrightarrow \text{no reaction} \quad \text{(Eq. 9)}$$
3° alcohol   (orange)

ALDEHYDES AND KETONES

Aldehydes and ketones are both "carbonyl compounds," compounds that contain a C=O group. Other carbonyl compounds are carboxylic acids and esters (see below) and amides (see Experiment 13).

Aldehydes are easily oxidized to produce organic acids (Equations 3 and 10); ketones, however, can be oxidized only under extreme conditions. In this experiment, a dilute solution of potassium permanganate, $KMnO_4$, is used as the oxidizing agent to distinguish an aldehyde from a ketone. If a reaction occurs, the dissolved purple permanganate ion, MnO_4^-, is reduced to insoluble brown MnO_2.

$$3\ RCHO + 2\ KMnO_4 + H_2O \longrightarrow 3\ RCOOH + 2\ MnO_2 + 2\ KOH \quad \text{(Eq. 10)}$$
aldehyde   (purple)   carboxylic acid   (brown)

$$R_2CO + KMnO_4 \longrightarrow \text{no reaction} \quad \text{(Eq. 11)}$$
ketone   (purple)

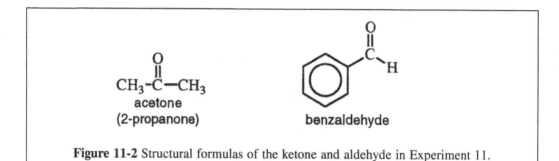

Figure 11-2 Structural formulas of the ketone and aldehyde in Experiment 11.

One test that can distinguish between alcohols and carbonyl compounds like aldehydes and ketones is the formation of yellow to red precipitates when the latter are put into an acidic solution of 2,4-dinitrophenylhydrazine. In Experiment 4, such products were used to illustrate the utility of thin layer chromatography in separating and identifying them.

Legal's test is one specifically for acetone and related compounds. In cases where abnormal quantities of lipid are metabolized, as in diabetes mellitus and starvation, an increase in the level of "ketone bodies" will appear in the urine. One of the ketone bodies, acetone, can readily be detected by adding an acid solution of sodium nitroprusside to the urine specimen. The appearance of a red color is a positive test for acetone.

CARBOXYLIC ACIDS

Carboxylic acids, RCOOH, are the classic weak acids. They have a polar region, the carboxyl group (—COOH), on a nonpolar chain. They are, as a rule, more soluble in water than are alcohols of equal carbon length because the carboxyl group is more polar than the hydroxyl group (—OH) of the alcohol. Carboxylic acids react readily with a base to form salts, which are usually water soluble.

$$\underset{\text{carboxylic acid}}{R-\overset{\overset{\displaystyle O}{\|}}{C}-O-H} + NaOH \longrightarrow \underset{\text{carboxylate salt}}{R-\overset{\overset{\displaystyle O}{\|}}{C}-O^-\ Na^+} +\ H_2O \qquad \text{(Eq. 12)}$$

$$\underset{\substack{\text{acetic acid}\\ \text{(ethanoic acid)}}}{CH_3-\overset{\overset{\displaystyle O}{\|}}{C}-OH} \qquad \underset{\text{propanoic acid}}{CH_3-CH_2-\overset{\overset{\displaystyle O}{\|}}{C}-OH} \qquad \underset{\text{pentanoic acid}}{CH_3CH_2CH_2CH_2-\overset{\overset{\displaystyle O}{\|}}{C}-OH}$$

$$\underset{\substack{\text{oxalic acid}\\ \text{(ethanedioic acid)}}}{HO-\overset{\overset{\displaystyle O}{\|}}{C}-\overset{\overset{\displaystyle O}{\|}}{C}-OH}$$

benzoic acid

salicylic acid
(2-hydroxybenzoic acid)

Figure 11-3 Structural formulas of carboxylic acids used in Experiment 11.

ESTERS

An ester, RCOOR', is structurally a derivative of a carboxylic acid. The hydrogen of the —OH is replaced by an alkyl or aryl (R) group, which may be different than the R of the acid. Esters are prepared by reacting a carboxylic acid with an alcohol in the presence of an acid catalyst, often sulfuric acid. Special methods can be used to remove water from the product side of the reaction, driving the equilibrium reaction to the right.

$$\underset{\text{carboxylic acid}}{R-\overset{\overset{\displaystyle O}{\|}}{C}-O-H} + \underset{\text{alcohol}}{H-O-R'} \xrightarrow{H_2SO_4} \underset{\text{ester}}{R-\overset{\overset{\displaystyle O}{\|}}{C}-O-R'} + \underset{\text{water}}{H-O-H} \qquad \text{(Eq. 13)}$$

Esters have pleasant odors, many of which are associated with flowers and various fruits. One of the esters to be made is methyl salicylate, which is an oil found in wintergreen and other plants and which finds use in liniments. See also Experiment 11.

ACETALS

Another very important class of oxygen-containing compounds are the hemiacetals and acetals. They are formed from aldehyde or ketone carbonyl (C=O) groups reacting with one or two hydroxyl groups, respectively.

hemiacetal

acetal

Hemiacetals are usually unstable, but with sugars they often prove stable and are a very important species. By reaction with a second hydroxyl group they go on to form acetals. Acetals are very stable in basic media but hydrolyze in acidic. The flavor molecules of vanilla beans and cinnamon are present as acetals, in which the second hydroxyl group is provided by glucose. Starch and cellulose are substances that are chains of glucose held together by acetal links.

Procedure

> **CAUTION: Many compounds of chromium are suspect carcinogens; handle sodium dichromate with care. Avoid contact.**
>
> **Potassium permanganate should be handled with care.**
>
> **Phenol is poisonous and caustic; do not handle with bare hands.**
>
> **Oxalic acid is poisonous.**

> Most organic compounds are flammable.
>
> Do not smell compounds unless directed to do so; physicians may advise pregnant women to avoid smelling any compounds in this experiment.
>
> Wash your hands after handling these and other substances.
>
> For handling chromium (VI) compounds, potassium permanganate, phenol, and oxalic acid, gloves should be worn.

I. PROPERTIES OF ALCOHOLS

1. Place 4 drops of the following alcohols in four numbered test tubes as follows:

   No. 1:   Ethanol
   No. 2:   1-Propanol
   No. 3:   1-Butanol
   No. 4:   1-Pentanol

2. Note the color and carefully check the odor of each alcohol by wafting the vapors toward your nose. The instructor will demonstrate this technique.

3. Add 3 mL of deionized water to each test tube; using a stirring rod, stir vigorously. Classify the compounds as insoluble (I) or soluble (S).

4. Review the use of universal indicator, pH paper, and litmus paper in Experiment 10.

5. With the stirring rod, apply a drop of any one of the water-soluble solutions (S) to wide-range pH indicator paper and record the pH, which will be the same for all.

6. Add a drop or two of universal indicator to each of the test tubes. Using the stirring rod, mix well! *Do not discard the solutions until you have completed Part II.*

II. PROPERTIES OF PHENOL

1. Put 1 mL of 5% aqueous phenol in a small test tube. **CAUTION: Phenol is a skin irritant; avoid skin contact! Wear gloves.**

2. Using the wafting technique demonstrated by the instructor, cautiously note the odor.

3. Using a stirring rod, test a drop or two of the solution on both red and blue litmus paper.

4. Add a drop of universal indicator to the solution. Stir well! Compare the color produced to those of the alcohol solutions of Part I, Step 5.

5. **Disposal:** Discard the solutions from Parts I and II into the "Waste Organics" container and wash the test tubes as directed by your instructor.

III. OXIDATION OF ALCOHOLS

1. Place 1 mL of reagent grade acetone (as a solvent) into each of four small test tubes. Label them 1-4.

2. To tube 1, add one drop of ethanol; to tube 2, add one drop of isopropyl alcohol; and to tube 3, add one drop of t-butyl alcohol. Do not add anything to the acetone in tube 4; it is a reference blank.

3. Working with one tube at a time, hold the tube firmly between two fingers near the top, add 2-4 drops of chromic acid reagent, and gently flick the tube at the bottom to mix it. DO NOT SHAKE IT; AVOID SKIN CONTACT. Observe changes that occur *in the first 2 seconds*. A green precipitate and loss of the orange color indicates that oxidation took place. (The chromic acid reagent used is a solution of chromic oxide in aqueous sulfuric acid.)

4. Record your results for each of the four samples.

5. **Disposal:** Discard the four test solutions into the "Waste Chromium Salts" container.

IV. ALDEHYDES AND KETONES—CLASSIFICATION TESTS

A. Distinguishing an Aldehyde from a Ketone

1. Pour about 1 mL of benzaldehyde into a test tube and try to identify its odor in one of the commercial products your instructor has available.

2. Add about 8 drops of 10% potassium permanganate solution, and stir vigorously. Note the color of the solution. Wash your hands.

3. Repeat Steps 1 and 2 with acetone.

4. **Disposal:** Discard the two solutions into the "Waste Manganese Dioxide" container.

B. Distinguishing Aldehydes and Ketones from Alcohols

1. Dissolve 2 drops of an unknown (record its sample number) to 2 mL of 95% ethanol and add 2 mL of 2,4-dinitrophenylhydrazine reagent.

2. Stir vigorously with a glass stirring rod and allow to stand for 10-15 minutes. Record whether or not a precipitate appears and what its color is.

3. **Disposal:** Discard tube contents in "Waste Organics" container.

C. Legal's Test for Acetone

1. Pour about 1 mL of acetone into a clean test tube and add 5 mL of deionized water.

2. Add a drop or two of 10% sodium hydroxide (NaOH) solution and then add several drops of a 5% solution of sodium nitroprusside. Observe the color of the solution. Wash your hands.

3. Repeat Legal's test (Steps 1 and 2) on commercial fingernail polish remover.

4. **Disposal:** Discard the two solutions into the "Waste Inorganics" container, or as directed by the instructor.

V. PROPERTIES OF SOME CARBOXYLIC ACIDS

1. Place the following carboxylic acids into four numbered test tubes:

   No. 1:   4 drops of propanoic acid
   No. 2:   4 drops of pentanoic acid
   No. 3:   0.1 g oxalic acid
   No. 4:   0.1 g benzoic acid

2. Carefully check the odor of each tube using the wafting technique.

3. Add 3 mL of deionized water to each test tube and stir vigorously. Classify the carboxylic acids as insoluble (I) or soluble (S).

4. Test 1 mL of each of the <u>water soluble (S) solutions</u> with 2 drops of universal indicator solution and record the pH.

5. To the remaining 2 mL of each of the <u>water soluble (S) solutions</u>, add 5 mL of 15% calcium chloride ($CaCl_2$) solution and stir vigorously. Indicate with a (+) on your Report Sheet if a precipitate is formed; use a (–) if none appears.

6. To each of the insoluble (I) carboxylic acids add 3 mL of 10% sodium hydroxide (NaOH) solution and stir vigorously. Use a (+) or (–) to indicate if a reaction has taken place. Wash your hands.

7. Using the benzoic acid/NaOH solution in tube No. 4, acidify it with 10% HCl. Check with test paper to be sure it is made acidic.

8. **Disposal:** Discard the solutions into the "Waste Organics" container or as directed by the instructor. Wash your hands.

VI. PREPARATION OF SOME ESTERS

1. Remove all flammable reagents from your working area. Using a 400-mL beaker, prepare a warm-water bath with temperature approximately 60°C.

2. Pour about 2 mL of 3-methyl-1-butanol into a test tube and add about 5 drops of concentrated sulfuric acid. **Handle concentrated sulfuric acid with caution. Immediately wash your hands.**

3. Add about 2 mL of acetic acid and place the test tube in the water bath for about 10 minutes. Note the odor produced using the wafting technique.

4. Repeat Steps 1 through 3 substituting the following alcohol/carboxylic acid combination:
   3 mL of methanol and 1 g of salicylic acid

5. Try to identify the esters you have synthesized by comparing their odors with those of commercial products.

6. **Disposal:** Discard the solutions into the "Waste Organics" container, or as directed by the instructor.

Name ________________________ Section ____________ Date ______________

Prelab Questions for Experiment 11

________ 1. Which of the following tests will be most helpful in diagnosing diabetes mellitus?
a. Measuring the specific gravity of the urine.
b. Treatment of urine with sodium nitroprusside reagent.
c. Heating the urine specimen.
d. Taking the pH of the urine.

2. Write structural formulas for the following compounds:

a. ethanol

b. acetone

c. acetic acid

d. isopropyl alcohol

e. methanol

3. Classify each of the following as a primary (1°), secondary (2°), or tertiary (3°) alcohol:

_____ a. $CH_3CH_2CH_2-OH$

_____ b. 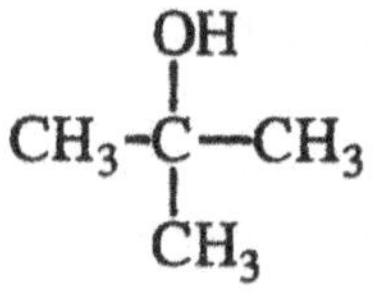 $CH_3-\overset{\displaystyle OH}{\underset{\displaystyle CH_3}{\overset{|}{\underset{|}{C}}}}-CH_3$

_____ c. $CH_3CH_2CH_2CH_2CH_2-OH$

_____ d. 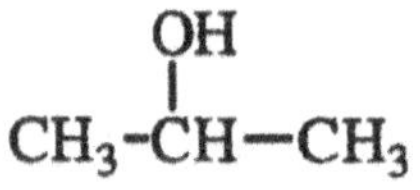 $CH_3-\overset{\displaystyle OH}{\overset{|}{CH}}-CH_3$

4. Water is a product in each of the following reactions. Complete each equation by writing the formula of water and the structural formula of the product.

a. $CH_3CH_2CH_2CH_2-OH \xrightarrow{[O]}$

b. 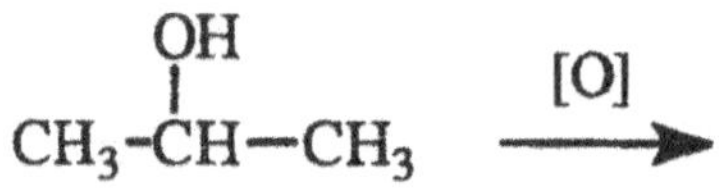 $CH_3-\overset{\displaystyle OH}{\overset{|}{CH}}-CH_3 \xrightarrow{[O]}$

c. 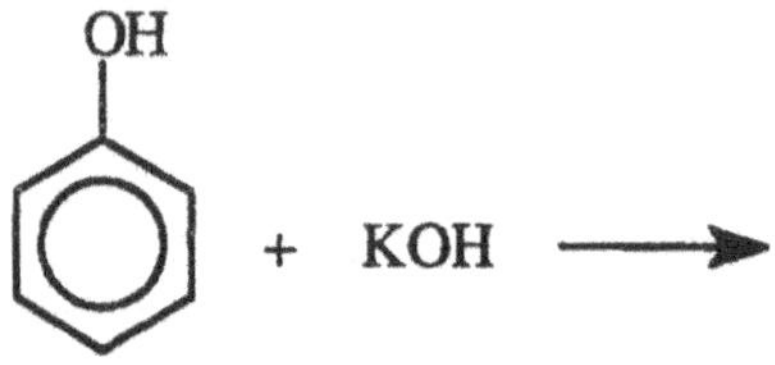 $+$ KOH $\longrightarrow$

d. 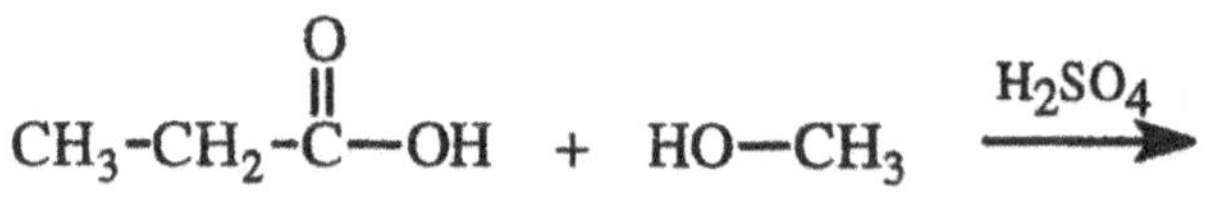 $CH_3-CH_2-\overset{\displaystyle O}{\overset{||}{C}}-OH$ $+$ HO$-CH_3 \xrightarrow{H_2SO_4}$

e. 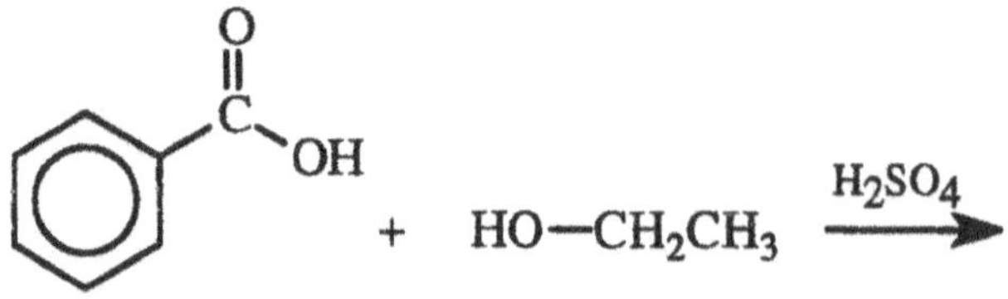 $+$ HO$-CH_2CH_3 \xrightarrow{H_2SO_4}$

Report for Experiment 11

I. PROPERTIES OF ALCOHOLS

    1. Data:

Compound	Structural Formula	Color	Odor (mild or acrid)	Solubility in H_2O (I or S)	pH
Ethanol					
1-Propanol					
1-Butanol					
1-Pentanol					

    2. a. What do your experimental results indicate concerning the relationship between the length of the carbon chain and the water solubility of alcohols?

         _____

        b. Explain the reason for this relationship.

         _____

    3. Alcohols (ROH) are classified in a distinct category apart from metal hydroxides such as NaOH. Explain how your data supports different categorizations.

         _____

II. PROPERTIES OF PHENOL

    1. Describe the odor of phenol. (Does its odor resemble that of any commercial product?)

         _____

         _____

    2. What was the pH of the phenol/water solution?

         _____

3. After adding a universal indicator, how did the phenol-water solution compare in color with the alcohol solutions in Part I?

_____

4. Phenols are in some respects similar to alcohols, but phenols and alcohols are placed in separate, distinct categories. Explain how your data supports different categorizations.

_____

III. OXIDATION OF ALCOHOLS

1. Data

Sample No.	Structural Formula of Solute	Color after Oxidation Test	Structural Formula of Product
1			
2			
3			
4			

2. Explain the result obtained when t-butyl alcohol was tested.

_____

_____

_____

IV. ALDEHYDES AND KETONES—CLASSIFICATION TESTS

A. Distinguishing an Aldehyde from a Ketone

1. a. Which of the commercial products has an odor similar to benzaldehyde?

_____

b. To acetone?

_____

2. a. Which of the compounds tested, benzaldehyde or acetone, reacted with 10% potassium permanganate?

_____

b. Where a reaction has taken place, write the structural formula for the organic product formed. Write "no reaction" if there was none.

B. Distinguishing Aldehydes and Ketones from Alcohols

1. Unknown's sample number: _____

2. Precipitate (Y or N)? _____

3. If yes, color? _____

4. Unknown is what type of compound? _____

C. Legal's Test for Acetone

Substance Tested	Color after adding sodium nitroprusside	Acetone Present? (+ or −)
Acetone		
Fingernail polish remover		

V. PROPERTIES OF SOME CARBOXYLIC ACIDS

1. Data

Compound	Structural Formula	Aroma (acrid or mild)	Solubility in Water (I or S)	Solubility in 10% NaOH (I or S)	pH	Reaction with $CaCl_2$ (+ or −)
Propanoic acid						
Pentanoic acid						
Oxalic acid						
Benzoic acid						

2. How do the solubilities of propanoic acid and pentanoic acid in water compare with the solubilities of alcohols of equal carbon length? Explain.

_____

_____

3. Explain the solubility results of benzoic acid with water and 10% NaOH.

_____

_____

4. Explain the result of acidifying the benzoic acid/NaOH mixture with HCl.

_____

_____

VI. PREPARATION OF SOME ESTERS

1. Write a complete balanced equation illustrating the preparation of esters from the following carboxylic acid/alcohol combinations:

   a. Acetic acid/3-methyl-l-butanol (also called isoamyl alcohol)

   b. Salicylic acid/methanol

2. Associate each of the above esters with a commercial product or fruit.

	Name of Ester	Product or Fruit
a.		
b.		

3. What is the function of the sulfuric acid in the above reactions?

_____

Organic Oxygen Compounds

Related Questions for Experiment 11

1. A student found an old bottle of benzaldehyde that had a poorly fitted lid. Although aldehydes are neutral compounds, the contents of the bottle turned blue litmus paper red. What has happened to the benzaldehyde? Hint: Benzaldehyde reacts easily with the oxygen in the air.

_____

_____

2. The labels have dried out and fallen off three bottles in the laboratory. Finding the labels, you know the compounds are propionaldehyde, acetone, and acetic acid. Design tests that can be run on a few drops of the chemical in each bottle to determine what compound is in each bottle. Explain the observations you would make with each test for each compound.

_____

_____

_____

_____

_____

_____

3. a. Write an equation for the neutralization of benzoic acid by sodium hydroxide.

   b. Write an equation for the reaction of hydrochloric acid with the organic product formed in Question 3a.

   c. The organic product of the reaction in Question 3a is sodium benzoate. Use *The Merck Index* to find the principal use of sodium benzoate.

_____

4. Use *The Merck Index* to find the principal use of calcium propionate (the calcium salt of propionic acid).

_____

5. a. Calcium ions are needed for blood coagulation. What were the results in Part V of treating calcium chloride with oxalic acid?

_____

   b. Explain why oxalic acid can be a deadly poison. (It occurs in low concentration in rhubarb.)

_____

Aspirin and Related Compounds

Reference: *General, Organic, and Biological Chemistry: An Integrated Approach, 4th ed., Chapter 8.*

Purpose: You will study salicylic acid and its medically useful ester derivatives, aspirin (acetylsalicylic acid) and methyl salicylate. This will include the stability and composition of aspirin tablets and also the saponification of either aspirin or methyl salicylate.

Introduction

Ancient Native Americans used the bark of the willow tree to counter fever and pain. Europeans learned of the medicinal properties of willow bark in 1763 when clergyman Edward Stone read a paper to the Royal Society of London. Willow bark extract was eventually found to be a powerful **analgesic** (pain reliever), **antipyretic** (fever reducer), and **anti-inflammatory** (reduces swelling) drug. The active ingredient in the bark of willow trees was found to be a compound that was named "salicin," derived from the botanical name *Salix* for willows. Salicin is an acetal formed from glucose and a compound that is metabolized to salicylic acid. Salicylic acid was easily made from phenol and carbon dioxide, but its use was limited because its acidic properties caused severe irritation of the stomach.

In 1893 the German chemist Felix Hoffman, working for the Bayer company, synthesized an ester of salicyclic acid by reacting its phenolic hydroxyl group with acetic acid, forming acetylsalicylic acid. It was marketed by Bayer under the trade name **aspirin**.

Like salicylic acid, aspirin is an analgesic, antipyretic, and anti-inflammatory; it is much less irritating to the stomach than salicylic acid. Aspirin does cause slight gastrointestinal bleeding that can, over time, cause iron deficiency or gastric ulcers. These complications can be avoided with enteric-coated aspirin, which does not dissolve until reaching the small intestine.

The standard adult aspirin tablet contains 325 mg (5 grains) of aspirin—the remainder of the tablet is made up of inert materials called binders, whose function is to hold the tablet together. Some manufacturers apply a micro-coating of hydroxypropyl methylcellulose to either make the pill easier to swallow or to delay dissolving until the tablet reaches the small intestine. Clinical studies among different brands of aspirin have shown no significant difference in relieving symptoms such as arthritis pain. Tablets containing aspirin sometimes include buffers to combat stomach irritation. Aspirin-containing combination pain relievers may also contain another analgesic and/or caffeine, a stimulant.

Aspirin is the most used medicine in the U.S.—more than 100 billion tablets are consumed each year for relief of headache, toothache, pain and fever of colds, muscle aches, menstrual pain, and pain of arthritis. Studies have suggested that one baby aspirin or one low dose aspirin tablet per day (both contain 81 mg of acetylsalicylic acid) may reduce the risk of heart attack, stroke, and colon cancer. Aspirin should not be given to children who have influenza or chicken pox because of the risk of the rare and often fatal Reye's Syndrome; children suffering from these diseases should be given a nonaspirin pain reliever such as acetaminophen. Unless directed by a physician, aspirin should not be taken during the last 3 months of pregnancy.

Only in the last few years have biochemists begun to understand how aspirin works its wonders. Aspirin prevents the formation of prostaglandins, a class of compounds responsible for evoking pain, fever, and local inflammation. Like many medicinal drugs, aspirin was developed from a naturally occurring substance. Chemists first isolated the active ingredient, determined its structure, and then improved on the original. Further improvement is possible once the exact mechanism of aspirin's interaction with prostaglandins is understood.

Salicylic acid is a bifunctional compound. In aspirin it is its hydroxyl group that is esterfied by reaction with acetic acid. In another derivative, methyl salicylate, which was made in Experiment 11, it is salicylic acid's carboxyl group that was esterified. That reaction with methanol afforded methyl salicylate. It, like salicin, from which the medicinal interest in salicylic acid arose, is also a naturally occurring compound; it is the principal ingredient of wintergreen oil, which can be distilled from the leaves of an evergreen herb, *Gaultheria procumbens*. One use is as a liniment. It is believed that its relief of muscle pain results from absorption through the skin and subsequent hydrolysis to salicylic acid.

Salicylic acid is a "common" name derived from the name of its original source; its systematic name is 2-hydroxybenzoic acid. The structural isomer 4-hydroxybenzoic acid also finds extensive use. A variety of its esters, formed by reaction of its carboxyl group with alcohols like methanol and butanol, are called *parabens* and are used as preservatives in cosmetics and pharmaceutics.

As mentioned, aspirin can be prepared in the laboratory by treating salicylic acid with acetic acid. Aspirin may be reconverted into salicylic acid in either of two ways. First, it may be hydrolyzed slowly in an acidic solution as illustrated in the following equation.

If aspirin is stored over a long period of time, especially under moist conditions, the above hydrolysis reaction takes place, and the characteristic odor of vinegar (acetic acid) is easily detected. Generally, "old" aspirin tablets contain traces of acetic acid and salicylic acid.

A second, and faster, method for converting aspirin into salicylic acid uses hydrolysis with sodium hydroxide (called "saponification," it is the method used to convert fats to soap), followed by acidification of the sodium salt.

Salicylic acid can be readily detected by the presence of a violet color upon addition of iron(III) chloride, and the presence of salicylic acid in an aspirin tablet is an indication of its decomposition.

Procedure

> **CAUTION: Methanol is flammable. Avoid breathing methanol vapor.**
>
> **CAUTION: Hydrochloric acid and sodium hydroxide can damage the skin; wash your hands after use.**

I. TEST FOR SALICYLIC ACID

1. Into one test tube, place 0.05 to 0.10 g of pure acetylsalicylic acid.

2. Into a second test tube, place 0.05 to 0.10 g of salicylic acid.

3. Add a few drops of 0.2 M iron(III) chloride, $FeCl_3$, to each of the two test tubes.

4. Dispose of the contents of the test tubes in the "Waste Inorganics" container.

II. ASPIRIN STABILITY AND SHELF LIFE

1. Obtain an "old" and a fresh aspirin tablet from your instructor. Check each tablet for a vinegar odor. Be sure to carefully waft the vapor toward your nose.

2. Place each tablet in a separate clean, dry test tube and grind it to a fine powder with a glass stirring rod. Be especially careful not to contaminate one tablet with another.

3. Add several drops of 0.2 M iron(III) chloride to each of the two powders. Note the color produced **immediately** after the reagent is added.

4. Dispose of the contents of the test tubes as in Part I.

III. ESTIMATING THE PERCENT OF ASPIRIN IN A TABLET

CAUTION: Methanol is flammable. Avoid breathing vapor of evaporated methanol.

1. Measure the mass of a 150-mL beaker.

2. Add 3 fresh aspirin tablets to the beaker; find the combined mass of beaker and tablets.

3. Pour 20 mL of methanol in the beaker.

4. Allow the aspirin to slowly disintegrate; break up the last few lumps with a stirring rod. The aspirin will dissolve in methanol, the binder will not. Stir for 3 minutes.

5. Determine the mass of a clean, dry evaporating dish.

6. Prepare a filtration setup with the weighed evaporating dish as the receiving vessel. (See Fundamental Laboratory Operations, Section 8) Stir the aspirin suspension vigorously and immediately filter it to remove the binder. Collect the filtrate in the weighed evaporating dish.

7. Pour an additional 5 mL of methanol in the beaker. Swirl the beaker to remove as much of the remaining aspirin as possible and quickly pour the suspension through the filtration setup of Step 6. Collect the filtrate in the same weighed evaporating dish. *Save the residue on the filter paper for use in Part IV.*

8. Perform this step in a fume hood. Carefully evaporate the filtrate to dryness by placing the evaporating dish over a hot-water bath. Avoid spattering! (See Fundamental Laboratory Operations, Section 5b.) **REMEMBER: Methanol is flammable and toxic.**

9. Allow the evaporating dish to cool, wipe off any moisture on the bottom of the dish, and measure its mass. Record the mass of aspirin. *Save the aspirin for use in Part V.*

IV. COMPOSITION OF THE BINDER

1. Transfer the residue on the filter paper to a test tube, add 5 mL of water; stopper and shake the tube.

2. Add a drop of 1% iodine solution.

3. Repeat Steps 1 and 2 with a pinch of soluble starch instead of the residue.

4. **Disposal:** Pour the contents of the test tubes down a sink drain.

V. SAPONIFICATION (HYDROLYSIS) OF ASPIRIN OR METHYL SALICYLATE

CAUTION: Hydrochloric acid and sodium hydroxide can damage the skin; wash your hands after use.

1. Carefully scrape the aspirin from Part III into a 100-mL beaker OR use 20 drops of methyl salicylate in the beaker.

2. Add 10 mL of 1 M NaOH, put a watch glass on the beaker, add a stir bar, and stir magnetically while heating on a hotplate until the mixture boils *gently*. Continue the *gentle heating* for 5 to 10 minutes. *Do not evaporate the solution to dryness*; if it approaches dryness, add enough water to bring the volume to above 5 mL. **CAUTION: Gloves should be worn.**

3. Allow the beaker to cool. Then add 4 drops of 6 M HCl (while stirring). Using the stirring rod, put a drop of the mixture on blue litmus paper to make sure it has become acidic.

4. Check to see if a precipitate has formed.

5. Add a few drops of 0.2 M iron(III) chloride to test for salicylic acid.

6. **Disposal:** Dispose of the contents of the beaker into the "Waste Inorganics" container.

Aspirin and Related Compounds

Prelab Questions for Experiment 12

_____ 1. The presence of salicylic acid in an aspirin tablet indicates
    a. the aspirin is fresh.
    b. the aspirin is old.
    c. the aspirin contains large quantities of binder.
    d. the aspirin contains small quantities of binder.

_____ 2. A sharp vinegar odor indicates
    a. the aspirin is fresh.
    b. the aspirin is old.
    c. the aspirin contains acetic acid as a binder.
    d. the aspirin has been buffered.

3. Given the following data, fill in the blank:

Mass of beaker and 3 tablets	75.42 g
Mass of beaker	74.28 g
Mass of 3 tablets	
Mass of evaporating dish and aspirin extract	42.48 g
Mass of evaporating dish	41.57 g
Mass of aspirin in 3 tablets	
Percent of aspirin in tablet	
Average mass of aspirin in 1 tablet	

Calculations:

4. a. What other chemical name is given to aspirin? _____

   b. Draw the structural formula of aspirin.

   c. Label the structural formula with the names of the functional groups (aldehyde, ketone, etc.).

5. a. Draw the structural formula of salicylic acid.

   b. Label the structural formula with the names of the functional groups (alcohol, ether, etc.).

   c. Which functional group reacts with acetic acid to form aspirin?

   _____

   d. Which functional group is responsible for causing stomach irritation?

   _____

6. a. Which functional group of salicylic acid reacts with methanol to form methyl salicylate?

   b. Draw the structural formula of methyl salicylate.

Report for Experiment 12

I. TEST FOR SALICYLIC ACID

Observations after addition of iron(III) chloride:

a. Pure acetylsalicylic acid _____

b. Pure salicylic acid _____

II. ASPIRIN STABILITY AND SHELF LIFE

1. Data:

	Old Aspirin	Fresh Aspirin
Odor		
$FeCl_3$ test		

2. What conclusions can you make regarding these results?

_____

3. What precautions should you take in storing aspirin tablets over a long period of time?

_____

4. Without using chemical reagents, how could you determine if aspirin is reasonably fresh?

_____

III. ESTIMATING THE PERCENT OF ASPIRIN IN A TABLET

1. Data:

Brand of aspirin	
Mass of beaker and 3 tablets	
Mass of beaker	
Mass of 3 tablets	
Mass of evaporating dish and aspirin extract	
Mass of evaporating dish	
Mass of aspirin in 3 tablets	
Percent of aspirin in tablet	
Average mass of aspirin in 1 tablet	

Calculations:

2. a. Do your experimental data indicate that the tested brand of aspirin tablet contains 5 grains of aspirin as claimed?

   _____

   b. Why might your data show an amount different than 5 grains?

   _____

   _____

IV. COMPOSITION OF BINDER

1. Results of test with iodine solution:

Brand of Aspirin	Results of Test with Iodine Solution

Starch	

2. What conclusions can you make about the binder composition in the brand of aspirin tested?

_____

_____

V. SAPONIFICATION OF ASPIRIN OR METHYL SALICYLATE

1. What visual change occurred in the reaction mixture on acidification?

2. Look up in *The Merck Index* the solubility of salicylic acid in water. How does this relate to the change on acidification?

3. Results of test with iron(III) chloride solution

4. What conclusion can be made concerning these results?

Aspirin and Related Compounds

Related Questions for Experiment 12

1. In most fatal cases of aspirin poisoning, the dosage has been greater than 20 grams. How many aspirin tablets are necessary to supply 20 grams of aspirin? _____

   Calculations:

2. The toxicity of a substance is usually given in the form "LD_{50} (method of dosage) in (test specie): x mg/kg (of body weight)." For example, for salicylic acid, *The Merck Index* gives: LD_{50} i.v. in mice: 500 mg/kg. "I.v." means intravenous, by injection into a vein.

   The LD_{50} must be known when prescribing medicines; to be effective, their doses often have to approach levels at which they may become toxic, and the physician has to know what that level is.

   If salicylic acid toxicity were the same for humans as for mice and a person weighs 132 lbs, what percentage would a 10 g dose be of the LD_{50} dose? [Show your calculations.]

3. Why do the textures of different brands of aspirin tablets differ?

   _____

   _____

4. Aspirin relieves the symptoms of head colds by reducing fever and easing aches and pains but does not shorten the time needed to "get over" the cold. Some physicians have suggested that aspirin may actually lengthen the period of a cold. Why might aspirin have this lengthening effect?

   _____

   _____

Analysis of an Aspirin Tablet

Organic Nitrogen Compounds

Reference: *General, Organic, and Biological Chemistry: An Integrated Approach, 4th ed., Chapter 8.*

Purpose: You will investigate the properties of amines and amides. You will prepare an amide that can identify the smell of improperly stored fish. Two typical alkaloids, quinine and caffeine, will be studied.

Introduction

AMINES AND AMIDES

Amines are derivatives of ammonia, NH_3, in which one or more hydrogen atoms of ammonia have been replaced by an organic group designated R. They are classified as primary, secondary, or tertiary, according to the number of R groups attached to the nitrogen.

$$
\begin{array}{cccc}
\text{H—N—H} & \text{R—N—H} & \text{R—N—H} & \text{R—N—R} \\
\mid & \mid & \mid & \mid \\
\text{H} & \text{H} & \text{R} & \text{R} \\
\text{ammonia} & 1° \text{ amine} & 2° \text{ amine} & 3° \text{ amine}
\end{array}
$$

Like ammonia, amines are basic compounds that react readily with acids to form salts. Here the reaction is with hydrochloric acid.

$$
\begin{array}{cccc}
\text{R—N—H} & + & \text{HCl} & \longrightarrow & \overset{\text{H}}{\underset{\text{H}}{\text{R—N}^{\oplus}\text{—H}}} \ \ Cl^{\ominus} \\
\mid & & & \\
\text{H} & & & \\
\text{amine} & & \text{acid} & & \text{amine salt} \\
\text{(base)} & & & \\
\end{array}
$$

The amine may be regenerated from the salt by treatment with a base such as sodium hydroxide.

$$
\overset{\text{H}}{\underset{\text{H}}{\text{R—N}^{\oplus}\text{—H}}} \ Cl^{\ominus} \ + \text{NaOH} \longrightarrow \text{R—N—H} \ + \ H_2O \ + \ \text{NaCl}
$$

amine salt   base   amine
(acid)   (base)

One of the amines used in this experiment has a bad smell like fish that have not been properly stored. It and other amines found in bad fish are thought to be why the practice of putting vinegar on fish arose before the era of modern refrigeration. This converts the amines to their acetate salts, which are not volatile and therefore diminish the smell. Some people have come to like the flavor of vinegar on fish and so the practice continues. The amine will be identified by conversion to an amide derived from benzoic acid.

Aniline is the simplest example of an aromatic amine, i.e., one that contains a phenyl group. It is used in manufacturing dyes and other commercial products. For example, its amide derived from acetic acid, acetanilide, has been used medicinally. See Figure 13-1 for structures of two amines used or discussed here.

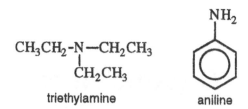

triethylamine   aniline

Figure 13-1 Structural formula of amines used or discussed in Experiment 13.

Amides are ammonia or amine derivatives of carboxylic acids. They may be classified as simple (nonsubstituted), monosubstituted, or disubstituted as indicated in the general formulas given below.

$$
\underset{\substack{\text{nonsubstituted}\\\text{amide}}}{R-\overset{\overset{\displaystyle O}{\|}}{C}-\overset{\displaystyle H}{\underset{\displaystyle H}{N}}-H}
\qquad
\underset{\substack{\text{monosubstituted}\\\text{amide}}}{R-\overset{\overset{\displaystyle O}{\|}}{C}-\overset{\displaystyle H}{\underset{\displaystyle R}{N}}-H}
\qquad
\underset{\substack{\text{disubstituted}\\\text{amide}}}{R-\overset{\overset{\displaystyle O}{\|}}{C}-\overset{\displaystyle R}{\underset{\displaystyle R}{N}}-R}
$$

The reactions of amines and amides studied in this experiment resemble reactions of proteins in body chemistry. In later studies we shall see that proteins are complex amides that are hydrolyzed in the process of digestion to amino acids—compounds containing both an amino ($-NH_2$) group and carboxyl ($-COOH$) group. The amino acids are absorbed into the bloodstream and transported to the cells, where they are reformed into proteins needed by the body. In summary:

$$
\underset{\substack{\text{proteins}\\\text{(amides)}}}{\text{proteins}\ +\ \text{water}} \xrightarrow{\text{(digestion)}} \underset{\substack{\text{amino acids}\\\text{(amine and carboxyl groups)}}}{\text{amino acids}} \xrightarrow{\text{(reactions in cells)}} \underset{\substack{\text{body proteins}\\\text{(amides)}}}{\text{body proteins}}
$$

Amides can be synthesized by several methods. You will synthesize one from the fish-derived amine and benzoyl chloride, an activated form of benzoic acid.

$$
\underset{\substack{\text{unknown}\\\text{amine}}}{R-\overset{\displaystyle R'}{\underset{}{N}}-H} \;+\; \underset{\substack{\text{benzoyl}\\\text{chloride}}}{Cl-\overset{\overset{\displaystyle O}{\|}}{C}-\bigcirc} \;\longrightarrow\; \underset{\substack{\text{benzamide of}\\\text{unknown amine}}}{R-\overset{\displaystyle R'}{\underset{}{N}}-\overset{\overset{\displaystyle O}{\|}}{C}-\bigcirc} \;+\; HCl
$$

Amides, like proteins, can be hydrolyzed but not as easily as esters. Amides can be broken down into carboxylic acids and ammonia (or amines) as illustrated in the following equations:

$$
R-\overset{\overset{\displaystyle O}{\|}}{C}-\overset{\displaystyle H}{\underset{\displaystyle H}{N}}-H \;+\; H-O-H \;\longrightarrow\; R-\overset{\overset{\displaystyle O}{\|}}{C}-OH \;+\; H-\overset{\displaystyle }{\underset{\displaystyle H}{N}}-H
$$

$$
R-\overset{\overset{\displaystyle O}{\|}}{C}-\overset{\displaystyle H}{\underset{\displaystyle R'}{N}}-H \;+\; H-O-H \;\longrightarrow\; R-\overset{\overset{\displaystyle O}{\|}}{C}-OH \;+\; H-\overset{\displaystyle }{\underset{\displaystyle R'}{N}}-H
$$

$$
R-\overset{\overset{\displaystyle O}{\|}}{C}-\overset{\displaystyle R'}{\underset{\displaystyle R''}{N}}-R' \;+\; H-O-H \;\longrightarrow\; R-\overset{\overset{\displaystyle O}{\|}}{C}-OH \;+\; H-\overset{\displaystyle }{\underset{\displaystyle R''}{N}}-R'
$$

Acetamide, when in its molten state, is a powerful solvent for many ionic and covalent compounds and is used in synthesizing other organic compounds. It commonly contains some impurity that gives it a mousy scent. An amide mentioned above, acetanilide, is used medicinally as an antipyretic and analgesic, alone and in combination with other drugs. Urea is the diamide of carbonic acid and is a waste product of protein metabolism; urea is removed from the blood by the kidneys and eliminated in the urine. Urea is used as a fertilizer and in urea-formaldehyde resins. Structures of these amides are shown in Figure 13-2.

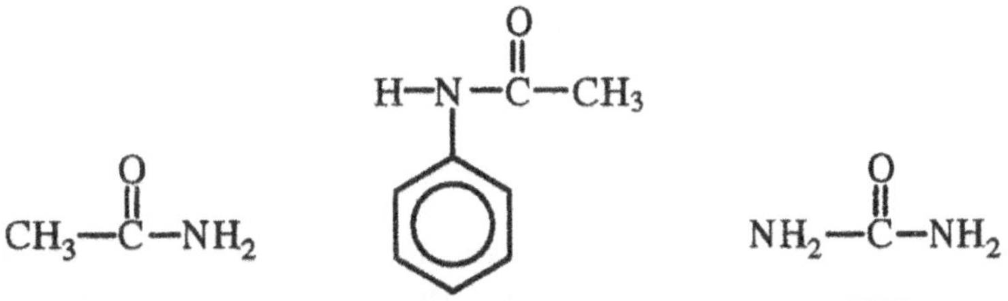

Figure 13-2 Structural formulas of amides used in Experiment 13.

ALKALOIDS

Alkaloids are naturally occurring nitrogen compounds having pronounced physiological activity. Like other simpler amines, alkaloids are bases; the name "alkaloid" comes from their alkaline properties. These complex molecules come from plants and many have medical uses. They generally are insoluble in water and are often found in commercial products in the form of their water-soluble acid salts.

$$\text{alkaloid} \quad + \quad \text{acid} \longrightarrow \text{alkaloid acid salt}$$
(water insoluble) (water soluble)

The pure alkaloid can be obtained as a precipitate from its acid salt by treatment with a base such as sodium hydroxide.

$$\text{alkaloid acid salt} \quad + \quad \text{base} \longrightarrow \text{alkaloid } (s)$$

Some of the better known alkaloids include atropine, used to dilate the pupils of the eyes; morphine, a painkiller; heroin, a synthetic modification of morphine that is a painkiller and narcotic; codeine, a painkiller and cough-suppressant; nicotine, a component of cigarette smoke and also a potent insecticide; reserpine, a tranquilizer; and cocaine, a local anesthetic and narcotic. The alkaloids you will study in this experiment, caffeine and quinine, were selected because of their ease of separation and their availability as components of commonly used foods.

The alkaloid **caffeine** is found in coffee, tea, and kola nuts. Caffeine is a mild stimulant to the central nervous system causing increased alertness and the ability to put off sleep. Caffeine is found in many over-the-counter medications because it tends to counteract the sedative effect of the active ingredient. The analgesic (pain-relieving) potency of aspirin may be increased by caffeine although some studies dispute this. Combination pain relievers often contain both aspirin and caffeine. Caffeine is the main ingredient of No-Doz® stay-awake tablets.

Coffee beans contain 2–5% caffeine, about the same amount present in tea leaves. Many people prefer to drink decaffeinated coffee. One of several solvents is used to extract the caffeine from the beans prior to roasting. There was once concern about the safety of the widely used solvent trichloroethylene; traces of the solvent remain in the decaffeinated coffee. The safer trichloroethane and the much safer liquid carbon dioxide are now used as solvents for the extraction.

Cola soft drink beverages are made from extract of the kola (cola) nut; the extract contains caffeine. Cola drink bottlers add phosphoric acid, caramel, sweeteners, and carbonated water. The U.S. Food and Drug Administration (FDA) requires that a "cola" contain some caffeine but limits the amount that can be present. Manufacturers usually remove all the caffeine from the kola extract and then add the correct amount. Caffeine is also added to some noncola soft drinks. Energy drinks are not included in FDA regulations and the caffeine limit is not disclosed on the labels. Some of these energy drinks have caffeine limits above the FDA limit imposed on cola or soda drinks.

Caffeine has little effect on the blood pressure in minute quantities but large amounts increase blood pressure. Individuals who drink coffee, tea, or cola drinks in large amounts can develop both a tolerance for and a dependence on caffeine. Heavy users can experience withdrawal symptoms of lethargy, headache, and even nausea after 18 hours of abstinence.

The structure of caffeine is very similar to that of adenine and guanine—important components of the genetic material DNA. This similarity has raised concern that caffeine might cause cancer or birth defects. So far there is little evidence to support this concern. Some people consider caffeine to be an addictive drug and some religions prohibit use of beverages containing caffeine.

caffeine   adenine   guanine

Quinine is an antipyretic (fever-reducer) and was for a long time the only known remedy for malaria. Quinine binds to the DNA of malaria-infected cells and inhibits their replication. Only infected cells are affected because they absorb quinine in higher concentrations than unaffected cells. The alkaloid is found in the bark of the cinchona tree; these trees were extensively cultivated in Indonesia in the late 19th century. When the Japanese invasion of Indonesia in World War II cut off the supply of quinine needed by Allied troops, American chemist Robert Burns Woodward succeeded in synthesizing quinine from coal tar. Woodward was renown for his ability to synthesize complex organic substances and was awarded the 1956 Nobel Prize in chemistry. Most alkaloids taste bitter, and quinine is frequently used as a standard reference for bitterness in taste studies. The bitter taste of tonic water is due to quinine, which is present as its sulfate salt. Water solutions of quinine *salts* are highly fluorescent, appearing light blue in the presence of ultraviolet light (blacklight).

Quinine

CAUTION:

1. Most organic compounds are flammable.

2. Handle all chemicals with care; avoid skin contact. Wash your hands after use.

3. Many amines have a pungent odor. Pregnant women should not check chemical odors.

Procedure

I. BASICITY OF AMINES

1. Place 5 drops of triethylamine in a test tube and add 2 mL of water. Observe the solubility of triethylamine. Test the solution with red and blue litmus paper. (Review the use of litmus paper in Experiment 10.)

2. **Cautiously** note the odor of the solution by gently fanning the vapor with your hand toward your nose.

3. Add 3 M hydrochloric acid to the solution drop-by-drop until the odor disappears. Write an equation for the reaction that has occurred.

4. Repeat Steps 1 through 3 with the "bad fish" amine to be used in Part II.

5. **Disposal:** Discard the solutions in the "Waste Amines" container.

II. IDENTIFICATION OF AMINE IN "BAD" FISH

CAUTION: Perform Steps 1-3 below in a hood. Wear gloves.

1. In a 16 × 150 mm test tube, put 20 drops of 10% NaOH solution and then add 5 drops of "bad" fish extract.

2. Add 10 drops of benzoyl chloride and then, using a glass stirring rod, *stir intermittently but vigorously for 5 to 10 minutes*.

3. Then add 9 drops of 3 M HCl with stirring. A precipitate should have formed; if not, scratch the inner wall of the test tube with the glass stirring rod.

4. Cool the reaction mixture in an ice bath for 5 to 10 minutes and then collect the crystalline product in a Buchner or Hirsch funnel by vacuum filtration. While the vacuum is still on, wash the crystals with 10 drops of water and then with 3 to 5 drops of ethanol. Leave the vacuum on for a few minutes to help dry the crystals.

5. **Disposal:** Pour the filtrate in the "Waste Acids" container.

6. Determine the melting point of the crystals as directed by your instructor. Then place the remainder of the dried product in a vial labeled with its melting point, the experiment number, and your name. Submit it to your instructor for evaluation. **Wash your gloved, then bare hands**.

III. THE HYDROLYSIS OF AMIDES

1. Add about 2 mL of 10% sodium hydroxide to a few crystals of acetamide in a test tube.

2. Warm the test tube in a hot water bath, and cautiously note the odor of the gas produced.

3. Place a strip of moist red litmus paper over the opening of the test tube. Reasoning from the structure of the amide used, name the odorous gas liberated.

4. Repeat Steps 1 through 3, substituting urea for acetamide.

5. **Disposal:** Discard the solutions in the "Waste Amines" container.

IV. PROPERTIES OF QUININE

1. Pour 25-mL portions of tonic water into each of two 150-mL beakers and label them samples A and B. Place the samples under blacklight (UV light) and observe their relative fluorescence.

2. Remove sample B from the blacklight and add, while stirring, 10% sodium hydroxide solution dropwise until a drop of the mixture turns red litmus paper blue. Then put the beaker back under the blacklight and compare its appearance relative to sample A.

3. Remove both samples from the blacklight and add, with stirring, 15 mL of deionized water to sample A and 15 mL of 0.1 M sulfuric acid to sample B.

4. Place both beakers under the blacklight and compare their appearance.

5. **Disposal:** Pour the contents of the beakers into the "Waste Acids" container.

V. PURIFICATION OF CAFFEINE FROM COFFEE

1. Put 1 g of crude caffeine extracted from coffee in a 150-mL beaker for purification by **sublimation.** This is the process of changing directly from a solid to the vapor state and back without ever being a liquid. [In Experiment 9, dry ice changed directly to gaseous carbon dioxide without liquefying but it doesn't resolidify without considerable cooling and pressure.] Here, the caffeine will sublime away from the less volatile impurities.

2. Place an evaporating dish filled with cold water on top of the beaker and heat the beaker <u>on low heat</u> on a hotplate. The crystals that form on the underside of the dish are purified caffeine. Only a few crystals need to be collected.

3. **Disposal:** Scrape the beaker residue and sublimed caffeine into the "Waste Amines" container.

Name ______________________ Section ____________ Date ______________

Prelab Questions for Experiment 13

_______ 1. What substances turn red litmus paper to a blue color?

a. acids
b. bases
c. alcohols
d. esters

2. Proteins belong to what class of organic compounds?

____

3. What are the hydrolysis products of a substituted amide?

____

4. If an amide produced ammonia upon hydrolysis, what would you conclude regarding the structure of the amide?

____

5. How does the preparation of an amide in Part II resemble the synthesis of body proteins?

____

6. List three physiological effects resulting from excessive intake of caffeine.

____

____

____

7. Aniline is insoluble in water but dissolves in dilute hydrochloric acid. Explain this (include the relevant equation).

Report for Experiment 13

I. BASICITY OF AMINES

1. Data:

Substance	Solubility in Water	Test with Litmus	Odor
Triethylamine			
"Bad Fish" amine			

2. Using structural formulas, write the equation for the reaction of triethylamine and hydrochloric acid.

3. Explain why the odor of the "bad fish" amine diminshed as hydrochloric acid was added.

II. IDENTIFICATION OF AMINE IN "BAD" FISH

1. What was the visual result of Steps 1–3?

2. Describe the appearance of the product on completion of Step 4.

3. What is the melting point of your dried product? _____
   Did it melt "sharply"—that is, within a narrow range?

4. The "bad fish" aroma is known to be from a small primary amine. The melting points of the benzamide derivatives of likely possibilities are:

Amine	M. P. of Benzamide (^{o}C)
Methanamine (CH_3NH_2)	80
Ethanamine ($CH_3CH_2NH_2$)	71
Propanamine ($CH_3CH_2CH_2NH_2$)	84
Butanamine ($CH_3CH_2CH_2CH_2NH_2$)	42

The presence of impurities lowers the observed melting point, and your product may not have dried completely. Observed melting points are never above reported ones, which are for extremely carefully purified samples. Allowing for that, which one or more of the above amines may be the one from bad fish?

III. THE HYDROLYSIS OF AMIDES

1. Using structural formulas, write equations for:

   a. the hydrolysis reaction of acetamide and water

   b. the hydrolysis reaction of urea and water

2. Did the red litmus paper change color when acetamide and urea were hydrolyzed? If so, what does it tell us?

IV. PROPERTIES OF QUININE

1. What compound caused the tonic water to be fluorescent?

   _____

2. Did the fluorescence diminish after addition of sodium hydroxide solution in Step 2? Why?

   _____

   _____

3. What reaction occurred when sodium hydroxide was added in Step 2? (A word equation is sufficient.)

4. a. How did the two solutions in Step 4 compare in fluorescence?

   _____

   b. Offer an explanation for this.

   _____

   _____

V. PURIFICATION OF CAFFEINE FROM COFFEE

1. Describe the appearance and odor of caffeine before and after sublimation.

Related Questions for Experiment 13

1. The analgesic property of acetanilide was discovered by accident when it was added by mistake to a patient's prescription. Acetanilide can be toxic; therefore scientists searched for more effective, less toxic variations. Two of the compounds tested, acetaminophen and phenacetin, were found to have effective analgesic and anti-pyretic properties but only aspirin reduced inflammation. Phenacetin was once an ingredient in APC (aspirin, phenacetin, caffeine) tablets but was removed from the market because it is a suspected carcinogen. Unlike aspirin, acetaminophen does not cause stomach irritation but overdoses can cause permanent liver damage.

   a. Draw the structural formula of acetanilide.

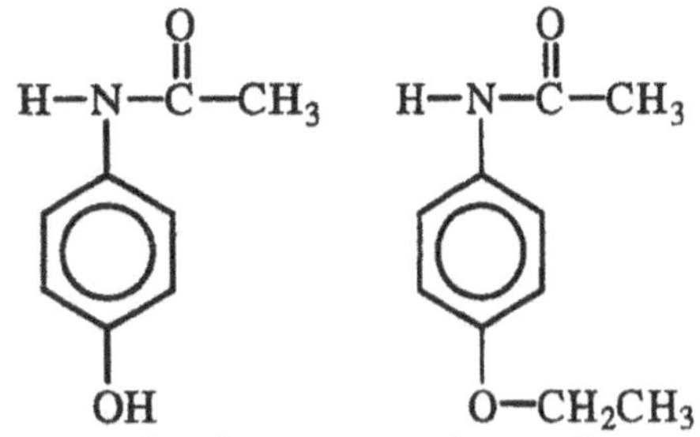

acetaminophen   phenacetin

   b. What chemical changes are necessary to convert acetanilide to acetaminophen?

   _____

   c. What common over-the-counter pain relievers contain acetaminophen?

   _____

   d. What chemical changes are necessary to convert acetanilide to phenacetin?

   _____

   e. Draw the structural formula of aspirin. (See Experiment 12.)

   f. What similarities exist in the structures of aspirin, acetaminophen, and phenacetin?

   _____

2. Caffeine stimulates the cerebral cortex by inhibiting an enzyme that in turn inactivates a form of the energy-supply molecule adenosine triphosphate (ATP). ATP is one of the most important molecules in living things; it is the immediate source of biochemical energy for powering reactions such as muscle contraction. ATP is also intimately involved with the metabolism of food and the biosynthesis of protein.

    a.   Write the structural formula of caffeine.

adenosine triphosphate (ATP)

    b.   In what way is the structure of caffeine similar to ATP?

_____

_____

3. The artificial sweetener aspartame (NutraSweet®) is a dipeptide—a very short protein molecule. The sweetness of aspartame was discovered accidentally in 1965 when a careless chemist licked his dirty fingers. He was fortunate that the compound was not toxic; his company was fortunate that he was curious to discover why his fingers were so sweet. Identify the functional groups (alcohol, amine, etc.) in the aspartame molecule.

    a.   _____

    b.   _____

    c.   _____

    d.   _____

    e.   Aspartame decomposes slowly in liquids, so soft drinks sweetened with it have limited shelf life. Which functional group is *easily* hydrolyzed?

_____

Experiment 14
Carbohydrates

Reference: *General, Organic, and Biological Chemistry: An Integrated Approach, 4th ed., Chapters 8, 9, and 10.*

Purpose: You will perform identification tests for sugars and polysaccharides. Carbohydrate digestion and absorption will be studied. Molecular models will be used to illustrate the structural differences between the α– and β–forms of D-glucose.

Introduction

Carbohydrates are a class of compounds that include the sugars (usually monosaccharides or disaccharides) and polysaccharides, chains of sugars linked together. Monosaccharides are polyhydroxy aldehydes like glucose or polyhydroxy ketones like fructose, whose structures are given below. Monosaccharides cannot be hydrolyzed into smaller units but polysaccharides can. Some examples are:

$$H_2O$$

A disaccharide (e.g., sucrose) $\longrightarrow$ two monosaccharide units

(*sometimes* different, as here: glucose and fructose)

$$xH_2O$$

A polysaccharide (e.g., starch) $\longrightarrow$ x (many) monosaccharide units

(*sometimes*, as here, all are the same kind: glucose)

Sugars have names ending in "-ose" and are classified by the number of carbon atoms in the molecule and also by the kind of carbonyl functional group present. For example, glucose is an aldohexose and fructose is a ketohexose.

OPTICAL STEREOISOMERISM

Another basic difference between sugars can be in the three-dimensional arrangements about their stereocenters. These are atoms that have four different atoms or groups attached, for example carbons 2 and 3 in this aldotetrose named D-erythrose:

$$
\begin{array}{c}
\overset{1}{\text{CHO}} \\
\text{H}\overset{2}{-\!\!-}\text{OH} \\
\text{H}\overset{3}{-\!\!-}\text{OH} \\
\overset{4}{\text{CH}_2\text{OH}}
\end{array}
\qquad
\begin{array}{c}
\text{CHO} \\
\text{H}\blacktriangleright\!\!-\!\!\blacktriangleleft\text{OH} \\
\text{H}\blacktriangleright\!\!-\!\!\blacktriangleleft\text{OH} \\
\text{CH}_2\text{OH}
\end{array}
$$

The chain is always numbered from the end that gives the lowest possible number to the carbonyl carbon.

Here, the structure of erythrose is drawn on the left as a Fischer projection, the simplest way to indicate the three-dimensional arrangement. But it only has three-dimensional meaning **IF** it is understood that all horizontal bonds project forward and all vertical bonds project backward, as shown specifically in the representation on the right. This always *has to be read into* a Fischer representation.

A "D" preceding the name of a sugar means that in its Fischer projection, the hydroxyl group on the highest-numbered stereocenter is on the right side. Virtually all common sugars have that arrangement and are said to be of the D-group. The rarely encountered isomer that is a mirror image would have its hydroxyl groups on the opposite side of the vertically drawn carbon chain of the Fischer projection and would be of the L-group. Here D comes from the Latin *dexter* (on the right side) and L from the Latin *laevus* (left). For example, the mirror images (called "enantiomers") of erythrose are:

D-erythrose   L-erythrose

And if the configuration is reversed at fewer than all stereocenters, here for example only at C.2, a stereoisomer is formed that is not a mirror image. It is called a diastereoisomer and is given a different name, here threose.

Its enantiomers, D- and L-threose, have these Fischer projection formulas:

D-threose   L-threose

These four aldotetrose isomers are the four predicted by the rule that the number of possible optical stereoisomers is equal to n^x, where x is the number of stereocenters (here, C.2 and C.3).

OPTICAL ACTIVITY

In polarized light, the light waves all lie in one plane. When polarized light passes through a solution of an optically active compound like D-glucose, the plane rotates. Each of a pair of enantiomers rotates the plane through the same number of degrees but in opposite directions. This rotation is a physical constant like a melting or boiling point, and it is reported in this form: α-D-glucose, immediately after being dissolved in water, has specific rotation $+112°$ (under designated conditions of temperature and wavelength) but it quickly changes (mutarotates) to $+52.7°$ because of the structural equilibration described below.

A mixture of equal concentrations of two enantiomers is called a "racemic mixture." It would show no optical rotation because the effects of the two enantiomers cancel each other. Another instance in which a material may contain stereocenters but not exhibit optical activity is where one half of the molecule is the mirror image of the other half. An example is this isomer of 2,3-butanediol:

It is symmetrical about this mirror plane.

THE STRUCTURES OF GLUCOSE AND FRUCTOSE

D-Glucose is the most commonly encountered sugar, probably because of its optimal three-dimensional structure. It has many aldohexose isomers, D-mannose and D-galactose being two of them. A ketohexose isomer is D-fructose, and D-glucose and D-fructose are the two monosaccharides formed by hydrolysis of table sugar, sucrose. The Fischer projection formulas for these two sugars are:

D-glucose   D-fructose

Note that they differ only in having their carbonyl groups interchanged between carbons 1 and 2.

In aqueous solution, most sugars exist predominantly in cyclic forms that are hemiacetals, compounds that have this structural feature: C-O-C-O-H. They are formed when an aldehyde or ketone undergoes reaction with a hydroxyl group:

As the arrows indicate, the equilibrium lies to the left, but when, as in the sugars, the alcohol hydroxyl group is in the same molecule as the carbonyl group, the equilibrium typically lies well to the right. With D-glucose, for example, hemiacetal forms comprise 98%. The equilibrium involves open-chain D-glucose and two cyclic hemiacetals (the bonds designating the hemiacetal feature are shown in **bold**):

α β

They are called "D-glucopyranoses," where the "pyran" means they contain a six-membered ring. Other sugar hemiacetals contain five-membered rings, designated by "furan." These are Haworth formulas, which depict the rings as being flat, and are drawn in the conventional way, with the ring oxygen to the back right. The dissolved molecules are, of course, always moving around, but drawing them in this standard way makes it easier to compare one sugar structure with another.

Notice that the two hemiacetals are labeled α and β. An α form has the C.1 hydroxyl on the opposite side of the ring from the CH$_2$OH group on C.5; the β isomer has these groups on the same side of the ring. The importance of this difference is illustrated by starch, which has its glucose monomer units α linked, and cellulose, in which they are β linked.

In Experiment 5, it was learned that six-membered rings are not flat; they are bent and exist almost entirely in the so-called chair form. This chair shape is shown in these more accurate depictions of the glucose hemiacetals:

α β

At equilibrium in aqueous solution, glucose is 36% α form, 2% open-chain, and 64% β form. It is the change to this equilibrium mixture on dissolving pure α-D-glucopyranose in water that accounts for the mutarotation discussed above, the optical rotation changing from +112° to +52°. Similarly, if β-D-glucopyranose (initial optical rotation +18.7°) is dissolved, equilibration occurs and brings the final optical rotation to the same +52°.

DI- AND POLYSACCHARIDES

These are derivatives of the simple sugars in which they serve as monomers that get linked together into polymer molecules called "di-," "tri-," "tetra-," etc., polysaccharides. The bonds that connect the sugar molecules are those of acetals. Hemiacetals have that name because they are halfway to acetals as shown in this pair of equations:

In the lab, hemiacetals can be isolated only if they spontaneously form and crystallize, as happens with many sugars. Acetals can be formed by introducing an acid catalyst and shifting the equilibrium to the right by removing one of the products, usually water. In living systems this is done enzymatically. There are two most important things to know about acetals. One is that the alcohol involved in the conversion of a sugar hemiacetal to an acetal can be that of a second sugar molecule; that is how the polysaccharides are assembled. The other important thing is that acetal links are stable in aqueous base BUT NOT in aqueous acid.

This explains why sucrose, an acetal that has has no free aldehyde or ketone group, does not give a positive Benedict's test, one that detects sugars that exist at least in equilibrium with a free aldehyde group; recall that ease of oxidation is a principal feature of aldehydes. This test requires alkaline conditions and therefore cannot release the aldehyde or ketone group from which an acetal is derived. Sucrose is an unusual disaccharide in that the carbonyl groups of *both* of its parent sugars are tied up in the acetal link.

D-glucose   +   D-fructose

sucrose

The bold bonds in the two sugars are those in their hemiacetal units; those in sucrose are just to indicate which edge of the rings projects toward the viewer.

While considering sugar-to-sugar acetal links, links of a type that are stable in basic media but not in acidic, the contrasting digestibilities of starch vs. cellulose needs comment. Humans can digest starch with its α glucose links easily but not cellulose with its β glucose links. Generally α and β acetal links are about equally easy to hydrolyze so the difference in these polyglucosides is surprising. It is parallel with the difference in mechanical properties of these two polymers. Starch has little if any strength when wet, but cellulose is the physical support of plants, much utilized in the form of wood. Cellulose polymer strands have a geometry that causes the strands to tightly stack up against each other, held by multiple hydrogen bonds between adjacent chains. This uniquely high stablitiy and the accompanying inaccessibility of the acetal units in cellulose explain its reluctant hydrolysis, even at the low pH of the stomach fluids.

SUGAR ALCOHOLS

The carbonyl group in aldoses and ketoses can be reduced with sodium borohydride ($NaBH_4$), by enzymes, or by hydrogen with platinum catalyst. This produces D-glucitol (also called "sorbitol") from D-glucose, for example.

$$
\begin{array}{ccc}
\text{CHO} & & \text{CH}_2\text{OH} \\
\text{H}\!-\!\text{OH} & & \text{H}\!-\!\text{OH} \\
\text{HO}\!-\!\text{H} & \xrightarrow{+2\,\text{H}} & \text{HO}\!-\!\text{H} \\
\text{H}\!-\!\text{OH} & & \text{H}\!-\!\text{OH} \\
\text{H}\!-\!\text{OH} & & \text{H}\!-\!\text{OH} \\
\text{CH}_2\text{OH} & & \text{CH}_2\text{OH} \\
\text{D-glucose} & & \text{D-glucitol}
\end{array}
$$

It is D-glucitol that accumulates in cataracts, degrading vision. Reduction of the aldopentose D-xylose produces xylitol, a sweetener recommended by dentists because mouth enzymes do not oxidize it to acids that cause cavities.

DIGESTION AND ABSORPTION OF CARBOHYDRATES

Before the cells of your body can utilize the energy stored in carbohydrates present in your diet, the carbohydrates must be digested and absorbed. **Digestion** is the process by which complex molecules are broken down into simple molecules. These simple molecules pass through the intestinal wall into the bloodstream during **absorption**.

The digestion of carbohydrates begins in the mouth as teeth tear food into tiny pieces; smaller pieces have a greater surface area and will be digested faster. Saliva contains an enzyme (amylase) that begins the hydrolysis of starch (a large polysaccharide) to dextrins (small polysaccharides) and maltose (a disaccharide). After swallowing, the food enters the stomach where protein and fat digestion begin but carbohydrate digestion temporarily ceases; the low pH of the stomach's gastric juice inactivates the salivary enzymes.

As food passes into the small intestine it is neutralized by alkaline pancreatic and intestinal juices. These juices also contain enzymes (maltase, lactase, sucrase) that renew the hydrolysis of complex carbohydrates. Eventually all polysaccharides and disaccharides are broken down to glucose, fructose, and galactose. These monosaccharides are small enough in size to pass through the intestinal wall and are absorbed into the blood. After circulating in the blood, fructose and galactose are converted into glucose by the liver. The glucose in the blood may be immediately used to provide energy for cellular activities or it may be stored as glycogen (a polysaccharide) in the liver and the muscles.

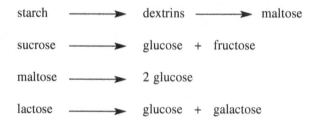

Starch and cellulose are the best known polysaccharides. Ordinary table sugar is the disaccharide sucrose. The digestion of starch and sucrose will be investigated in Part V of this experiment.

Absorption is a form of **dialysis**—the movement of small molecules through a membrane. In Part I of this experiment a solution of starch and glucose will be placed inside a cellophane dialysis bag. After the dialysis bag has been immersed in water for a length of time, the water will be tested for the presence of starch and glucose.

TESTS FOR CARBOHYDRATES

A. Molisch's Test

This is a test for any carbohydrate. A small quantity of solution to be tested is poured into a test tube, and an alcoholic solution of α-naphthol is added. The tube is then held at an angle while cold, concentrated sulfuric acid is slowly poured down the side. A purplish color indicates a positive test. Note: Other compounds such as glycoproteins can give a positive test since they can be hydrolyzed by a strong mineral acid such as sulfuric acid, releasing monosaccharides.

B. Seliwanoff's Test

This test is used to distinguish a ketose from an aldose. The test solution is mixed with hot hydrochloric acid and resorcinol (Seliwanoff's reagent). Ketohexoses will produce a red color; aldohexoses will produce a pink color. The test is of little value when performed on a disaccharide or polysaccharide since aqueous hydrochloric acid will hydrolyze them to mixtures of monosaccharides, which may be mixtures of ketoses and aldoses.

C. Tests for a Reducing Sugar

These tests indicate the presence of an aldehyde group. The general reaction is:

$$Cu^{2+} + \text{reducing sugar} \xrightarrow{\text{heat, base}} Cu_2O + \text{oxidized sugar}$$

(blue)   (free aldehyde)   (brick red)

The reagent used is called "Benedict's reagent," but many modifications, such as the Clinitest reagent, are frequently substituted. All <u>mono</u>saccharides will provide a positive Benedict's test, even those that do not contain an aldehyde group. Monosaccharides that are ketoses rearrange their structures to form an aldehyde group under test conditions and, therefore, give a positive test. Only those disaccharides having a free aldehyde group will react with the test reagent; sucrose is the only common disaccharide that does not react. Sugars that react are **reducing sugars***; those that are unreactive are **nonreducing sugars**. Sucrose does not react because it is an acetal and therefore stable in base.

D. Test for a Polysaccharide

A dilute solution of iodine will produce the following colors when mixed with some polysaccharides, e.g., starch—blue; dextrins—red; disaccharides and monosaccharides—colorless. Iodine solutions will not give an identifying color with cellulose.

---

*Copper(II) ion, Cu^{2+}, of Benedict's reagent is **reduced** to copper(I) ion, Cu^+, in the red product Cu_2O. Sugars that reduce Cu^{2+} are **reducing sugars**. The aldehyde group of the sugar gains an oxygen atom to become a carboxylic acid; the reducing sugar is **oxidized** in the course of the reaction.

Procedure

I. ABSORPTION OF CARBOHYDRATES

1. Cut a 15-cm (6-inch) length of dialysis tubing. With a 4-inch piece of string, **tightly** tie it 1 inch from one end.

2. Half fill the tubing with a 1% cooked starch solution; then add 20% glucose solution until full. **Tightly** tie the tubing with a piece of string 1 inch from the second end. Quicky rinse the bag.

3. Using a stirring rod and the strings on the dialysis bag, suspend it in a 400-mL beaker of deionized water as shown in Figure 14-1.

Figure 14-1 Dialysis bag suspended in a beaker of water.

4. Allow the bag to remain in the water for about 90 minutes while you proceed to later Parts II to IV of the Experiment.

5. After the 90 minutes, remove the bag (SAVE the surrounding liquid for the testing below), cut the string at one end, and pour the contents into a 150-mL beaker.

6. Test 1 mL of each of the two solutions with Benedict's reagent as in Part IIC. A positive test indicates the presence of glucose.

7. Test 1 mL of the solution inside the bag and 1 mL of the solution outside the bag with iodine solution as in Part IIE. A positive test indicates the presence of starch.

8. Pour remaining contents of beaker and dialysis bag into any sink. Dispose of test mixtures as in Part II.

II. TESTS FOR CARBOHYDRATES

To facilitate performing the tests of Parts A-E on each of the following solutions, pour about 10 mL of each into numbered test tubes:

No. 1:   1% D-glucose
No. 2:   1% D-fructose
No. 3:   1% sucrose
No. 4:   1% uncooked starch

From these tubes, use a fresh portion for each test.

On completion of each test, DISPOSE of the test mixture in the "Waste Organics" container.

A. Molisch's Test (general for carbohydrates)

CAUTION: Handle the high-concentration sulfuric acid with care; keep it and the test reagent off skin and wash after using.

1. Place 10 drops of the solution to be tested into a test tube. While stirring with a glass-rod, add 5 mL of room-temperature 75% sulfuric acid (H_2SO_4; prepared in advance by your instructor).

2. Add 3 drops of Molisch's reagent (3% α-naphthol in ethanol) and again stir thoroughly. The mixture becomes yellow.

3. Warm the test tube to about 80 °C in a hot water bath. If the mixture becomes red to violet in color, the presence of a carbohydrate is indicated.

B. Seliwanoff's Test (for ketoses)

CAUTION: Handle this resorcinol/hydrochloric acid reagent with care and keep it off skin; wash after using.

1. Put 20 drops of the solution to be tested into a test tube and add 4 mL of Seliwanoff's reagent (0.15% resorcinol in 3M HCl).

2. Place the test tube in a boiling water bath for about 8 minutes.

3. If the mixture becomes bright red in color, the presence of a ketose is indicated. (The warm, acidic conditions hydrolyze acetals.)

C. Benedict's Test (for reducing sugars)

1. Put 20 drops of the solution to be tested into a test tube and add 5 mL of Benedict's solution [copper(II) sulfate in aqueous sodium citrate/sodium carbonate].

2. Place the test tube in a boiling water bath for at least 5 minutes.

3. Remove the test tube and observe whether or not a color change occurred. A change to a murky red-orange indicates reduction of the dissolved copper(II) ions to insoluble copper(I) oxide by a reducing sugar.

D. Clinitest Method (for reducing sugars)

CAUTION: Clinitest tablets contain the strong base sodium hydroxide. Avoid contact with skin, eyes, mucous membranes, and clothing. Wash after using.

1. Put 20 drops of the solution to be tested into a test tube and drop a Clinitest tablet into it. *Do not handle the tablet with bare fingers.*

2. After 15 seconds, compare the color of the solution to the color chart.

E. Iodine Test (for starch and dextrins)

1. Put 20 drops of the solution to be tested into a test tube and add a drop or two of 1% iodine solution.

2. Observe the color produced. A blue or blue-black color indicates presence of starch; a red color indicates presence of dextrins, partially hydrolyzed starch.

III. THE α- AND β-FORMS OF D-GLUCOPYRANOSE

Use the following instructions and a provided model set to make models of D-glucose in its open-chain form and in both of its α- and β-pyranose hemiacetal forms. *Include* the hydroxyl (OH) groups in your models.

Build the open chain form first. Work from the Fischer projection formula shown in the Introduction. (Keep in mind the numbering used for glucose: C.1 is the carbon of the aldehyde group and the other carbons of the chain are designated sequentially as C.2, C.3, etc.) *Have your instructor note approval of your open-chain model on your report form.*

Now build models of both the α- and β-forms of D-glucopyranose. It doesn't matter which one first. To make one of the hemiacetal forms, use the oxygen of the hydroxyl group at C.5 of the open-chain model to make a ring with the aldehyde carbon, C.1. You will need to break and make some bonds. Remove the hydrogen from the oxygen at C.5. Open the double bond of the aldehyde group. Bond the oxygen at C.5 to C.1, and change the former aldehyde oxygen to a hydroxyl group by placing the hydrogen that used to be on the oxygen at C.5 on what was the aldehyde oxygen. The result should be a hemiacetal functional group at C.1. If what you built has the hemiacetal hydroxyl group *trans* on the ring to C.6 (in the CH_2OH group), you have made the α form of D-glucose. If it is *cis*, you have made the β form. (This is the standard method for distinguishing α- and β-hemiacetals.) Consult your textbook if you have questions. *Get your instructor's initials in the report section for both of the cyclic models.*

IV. DIGESTION OF CARBOHYDRATES

A. Enzymatic Hydrolysis of Starch

A 1% solution of commercial α-amylase from human saliva (free from reducing sugars) will be used here as the "saliva" sample.

To test in the absence of enzyme (as a control):

1. With a mortar and pestle, grind up a soda cracker and transfer it to a 16 × 125-mm test tube.

2. Add 10 mL of water and place the tube in a 35–40 °C water bath for 10 minutes. Stir frequently.

3. Filter the suspension, using vacuum filtration.

4. Test 1 mL of the filtrate for glucose as in Part IIC and another 1-mL portion for starch as in Part IIE.

To test the effect of α-amylase, do this parallel experiment:

1. In the mortar and pestle, grind up another soda cracker and transfer it to a 16 × 125-mm test tube.

2. Add 10 mL of the "saliva" solution and place the tube in the 35–40 °C water bath for 10 minutes. Stir frequently.

3. Filter the suspension, using vacuum filtration; the equipment used above can be reused after rinsing and subsequent replacement of the filter paper.

4. Test 1 mL of the filtrate for glucose as in Part IIC and another 1-mL portion for starch as in Part IIE.

Disposal: Pour remaining cracker suspensions in any sink and the test mixtures into "Waste Organics" container.

B. Acid-Catalyzed Hydrolysis of Sucrose

1. Pour 10 mL of a 1% sucrose solution into a test tube.

2. Add 1 mL of 3 M HCl (**CAUTION!**) and place the test tube in a boiling-water bath over a hot plate for 10 minutes.

3. Pour approximately 1 mL of the solution into a second test tube and insert a strip of red litmus paper.

4. Neutralize the HCl by adding 10% sodium bicarbonate solution drop-by-drop until the red litmus just turns blue.

5. Test 1 mL of the solution with Benedict's reagent as in Part IIC.

6. **Disposal:** Same as in Part IVA.

Carbohydrates

Prelab Questions for Experiment 14

_____ 1. Which of the following compounds will give a positive Molisch test?
a. maltose
b. galactose
c. cellulose
d. all of the above (a, b, and c)
e. none of the above (a, b, and c)

_____ 2. Which of the following statements explains the basis for obtaining a positive Benedict's test?
a. Secondary alcohols can be oxidized to form ketones.
b. Primary alcohols can be oxidized to form aldehydes.
c. Aldehydes are easily oxidized to form carboxylic acids.
d. Tertiary alcohols cannot be oxidized easily.

_____ 3. All carbohydrates are classified according to the
a. compound(s) produced upon hydrolysis.
b. number of monosaccharides produced upon hydrolysis.
c. functional group they contain.
d. all of the above

_____ 4. Seliwanoff's test is used to distinguish
a. a ketose from an aldose.
b. a monosaccharide from a disaccharide.
c. a monosaccharide from a polysaccharide.
d. a reducing sugar from a nonreducing sugar.

_____ 5. When a dilute iodine solution is placed on a slice of bread, the solution turns dark blue because bread contains
a. glucose.
b. sucrose.
c. starch.
d. protein.

_____ 6. Sucrose is a nonreducing sugar because
a. it makes a person gain weight.
b. it does not react with Cu^{2+}.
c. it is a monosaccharide.
d. it is a disaccharide.

7. Name the monosaccharides formed in the hydrolysis of:

a. starch _____

b. sucrose _____

8. What part of the body does the dialysis bag in Figure 14-1 represent? _____

Carbohydrates

Report for Experiment 14

I. ABSORPTION OF CARBOHYDRATES

1. Data. Use a (+) to indicate a positive test and a (−) to indicate a negative test:

	Iodine Solution	*Benedict's Test*
Solution inside bag		
Solution outside bag		

2. Explain why you obtained the above results.

_____

_____

_____

II. TESTS FOR CARBOHYDRATES

1. Data:

Carbohydrate Solution	*Molisch (+ or −)*	*Seliwanoff's (Red or Pink)*	*Benedict's (+ or −)*	*Clinitest (%)*	*Iodine (+ or −)*
1% D-glucose					
1% D-fructose					
1% sucrose					
1% uncooked starch					

2. Explain your results of the Seliwanoff's test with sucrose solution.

_____

_____

3. A student tested starch with Benedict's solution and obtained a weak positive test. Explain.

_____

_____

III. THE α- AND β-FORMS OF D-GLUCOPYRANOSE

1. Instructor's approval of model of open-chain D-glucose:

   _____

2. Sketches of the two models of the two cyclic structures.

3. Instructor's approval of the models of the D-glucopyranoses:

   _____

4. What class of compound is formed when the open-chain structure of D-glucose converts to the cyclic form?

   _____

5. What kind of isomers, conformational or geometric, are α-glucose and β-glucose?

   _____

IV. DIGESTION OF CARBOHYDRATES

A. *Enzymatic Hydrolysis of Starch*

1. Data. Indicate the results of these tests on crushed crackers:

	Iodine Test	*Benedict's Test*
No enzyme		
Enzyme-treated		

2. Explain the test results.

   Control (no enzyme) _____

   _____

   Enzyme-treated _____

   _____

3. Write a word equation to illustrate the reaction that occurs when starch is completely hydrolyzed.

   _____

B. *Acid-Catalyzed Hydrolysis of Sucrose*

1. Data. Indicate the results of these tests:

	Benedict's Test
Sucrose (results from Part IIC)	
Sucrose after being hydrolyzed in HCl	

2. Explain why hydrolyzed sucrose gave different results from unhydrolyzed sucrose.

   _____

   _____

3. Write Fischer projection formulas for the products formed when sucrose is hydrolyzed.

Related Questions for Experiment 14

1. After chewing on a piece of toast for five minutes, a student noticed a sweet taste in her mouth. Explain her observations.

   _____

   _____

2. Why is D-glucose (dextrose), rather than sucrose, given intravenously to patients?

   _____

   _____

3. Glucose in the urine may be a symptom of what disease?

   _____

4. Explain why infant formulas often contain mixtures of dextrins and maltose rather than starch.

   _____

   _____

5. a. What polysaccharide consists of α-D-glucose units? _____

   b. What polysaccharide consists of β-D-glucose units? _____

6. List two advantages in using the Clinitest reagent in preference to Benedict's reagent.

   _____

   _____

7. Why is it necessary that you digest (hydrolyze) starchy food?

   _____

8. D-Xylose, on completion of mutarotation, has specific rotation +18.6°. When its aldehyde group is reduced to a –CH$_2$OH group, what will the specific rotation of the product be?

$$
\begin{array}{c}
\text{CHO} \\
\text{H} \!-\!\!\!-\!\!\!- \text{OH} \\
\text{HO} \!-\!\!\!-\!\!\!- \text{H} \\
\text{H} \!-\!\!\!-\!\!\!- \text{OH} \\
\text{CH}_2\text{OH}
\end{array}
$$

D-xylose

Carbohydrates

Lipids

Reference: General, Organic, and Biological Chemistry: An Integrated Approach, 4th ed., Chapters 8, 9, and 11.

Purpose: You will investigate physical and chemical properties of some lipids including solubility, identification tests, and tests for rancidity and unsaturation. You will compare the action of soap to a detergent in hard and soft water.

Introduction

Lipids are substances that may be extracted from plant or animal tissues using nonpolar solvents. They are grouped into two major classes: those that can be saponified (hydrolyzed by a base) and those that cannot.

SAPONIFIABLE LIPIDS

The saponifiable lipids that you will study here are triglycerides,* triesters of glycerol (1,2,3-propanetriol) and three, rarely identical, fatty acids. These are carboxylic acids (RCOOH) that have large R groups (typically containing 11 or more carbon atoms) and of varyingly complex structure. Their total carbon count is usually an even number as they are biosynthesized from acetic acid (C_2) units. Other types of saponifiable lipids may simply have fewer ester groups (the mono- and diglycerides) or may have one fatty acid unit replaced with a sugar or phosphoric acid unit, which may in turn be linked to other units as in the glycolipids and phospholipids. Triglycerides from plants tend to be liquids and are called **oils**; **fats** are solid triglycerides from animals. **Saturated fatty acids** have only single bonds between carbon atoms. **Unsaturated fatty acids** contain carbon-to-carbon double bonds. Unsaturated fatty acids with one carbon-carbon double bond are **monounsaturated**; those with two or more carbon-carbon double bonds are **polyunsaturated**. The fatty acids in animal fat tend to be saturated fatty acids; examples include the fats in beef, chicken, butter, and lard. Fatty acids in vegetable oils have a higher degree of unsaturation but this varies considerably from vegetable to vegetable. Tropical oils such as coconut oil and palm oil are high in saturated fatty acids. Olive oil and canola (rapeseed) oil are high in monounsaturated fatty acids while safflower oil, sunflower oil, and corn oil are high in polyunsaturated fatty acids. In this experiment you will investigate cottonseed oil (about 25% saturated, 25% monounsaturated, and 50% polyunsaturated fatty acids) and other commercial vegetable oils.

A characteristic feature of glycerides is that when they, e.g., drip onto hot charcoal, they degrade in part to the aldehyde acrolein, $H_2C=CHCHO$, a mildly toxic substance that has an easily recognizable odor. It forms from the glycerol portion of the fat, and its formation can be catalyzed by an acidic substance like potassium hydrogen sulfate. Production of the distinctive odor of acrolein can serve as a test for glycerides.

Much remains unknown about the health effects of dietary fat, but current research indicates that Americans consume far too much. The amounts of triglycerides and cholesterol found in the blood

---

*Triglycerides are also called **triacylglycerides**, **triacylglycerols**, or **glyceryl trialkanoates**.

are useful in assessing a person's risk of cardiovascular (heart and artery) disease, and most physicians recommend reducing the amount of fat in the diet.

Margarine and solid shortening (such as Crisco®) are made by hydrogenation of unsaturated vegetable oils. The addition of hydrogen to carbon-carbon double bonds raises the melting point of the triglyceride; where vegetable oils are liquid, hydrogenated vegetable oils are solid. Natural vegetable oils contain only *cis-* double bonds. Partially hydrogenated margarine and shortening contain up to 40 percent of the *trans-* isomer. Recent research suggests the *trans-* isomers of margarine and shortening are linked to lower HDL (High-Density Lipids) levels and may increase the rate of heart disease.

People need *some* fat in their diet; linoleic acid and linolenic acid are essential fatty acids that human beings cannot synthesize. These fatty acids or their derivatives help regulate the formation of blood clots, blood pressure, the immune response, the inflammation response to injury, and many other body functions. Linoleic acid and linolenic acid are polyunsaturated fatty acids; both are widely distributed in foods and are easily stored in the body, so deficiencies are unlikely.

The degree of relative unsaturation of fats and oils can be compared by measuring the quantity of bromine that will react with equal amounts of lipids. Recall the test for unsaturation performed in Experiment 5:

a typical alkene
(colorless)   (red-brown)   (colorless)

Oils are more easily metabolized than fats, but they become rancid more quickly because of the ease with which they are oxidized. A hydrogen on a carbon bonded to a carbon-carbon double bond, and especially ones attached to two double bonds (as in linoleic acid), very readily react with oxygen to produce hydroperoxides.

$$CH_3(CH_2)_4CH=CHCH_2CH=CH(CH_2)_7CO_2H + O_2 \longrightarrow$$

And these lead to the break up of the carbon chain into smaller, more volatile products that are unpleasant smelling aldehydes and acids, explaining the bad odor of rancid foods. In this experiment, a test will be made for the presence of hydroperoxides in an old sample of a vegetable oil and another (the Kreiss test) for the presence of resulting breakup products.

Hydrolysis (splitting apart by the addition of water) of a fat or oil produces glycerol and three fatty acids.

$$\text{fat or oil} + 3\ H_2O \longrightarrow \text{glycerol} + 3\ \text{fatty acids}$$

$$H-\underset{\underset{H}{\overset{\overset{H}{|}}{|}}}{C}-O-\overset{\overset{O}{\parallel}}{C}-R \ + \ H-O-H$$

$$H-\overset{\overset{O}{\parallel}}{\underset{|}{C}}-O-\overset{\overset{O}{\parallel}}{C}-R \ + \ H-O-H \quad \longrightarrow$$

$$H-\underset{\underset{H}{\overset{|}{}}}{C}-O-\overset{\overset{O}{\parallel}}{C}-R \ + \ H-O-H$$

triglyceride　　　　　　water

$$H-\underset{\underset{H}{\overset{\overset{H}{|}}{|}}}{C}-O-H \ + \ H-O-\overset{\overset{O}{\parallel}}{C}-R$$

$$H-\underset{|}{C}-O-H \ + \ H-O-\overset{\overset{O}{\parallel}}{C}-R$$

$$H-\underset{\underset{H}{\overset{|}{}}}{C}-O-H \ + \ H-O-\overset{\overset{O}{\parallel}}{C}-R$$

glycerol　　　　　　fatty acids

When hydrolysis occurs in the presence of a base, soap and glycerol are formed. Soap is the salt of a fatty acid; hydrolysis by a strong base is called **saponification**. In Experiment 12 you saponified the ester group of aspirin.

When an alcohol like methanol is used instead of water and only a catalytic amount of base (or acid), a similar process occurs. It is called "transesterification" and converts the triglyceride to methyl esters of the same fatty acids. As in hydrolysis, glycerol is the byproduct. The resulting mixture of methyl esters is "biodiesel."

SOAPS AND DETERGENTS

A **soap** is the sodium or potassium salt of a fatty acid. The greater part of the soap molecule is nonpolar, but the polar "head" at the carboxyl end permits a small portion of the molecule to be water soluble. The cleansing power of soap is dependent on this water solubility. Water containing Ca^{2+} and Mg^{2+} ("hard" water) forms insoluble salts with soap and reduces its cleansing properties. Water softeners have ion-exchange resins that replace Ca^{2+} and Mg^{2+} with Na^+ which does not interfere with the action of soap.

The ion exchange resins used here are plastic beads usually made of polystyrene that has had sulfonic acid groups bonded to its polymer chains. These groups are derived from sulfuric acid and have its ability to be neutralized, converting them to salts. When ready for use, they are in their sodium salt form, and then they can exchange these sodium ions for the divalent (therefore more tightly held) calcium and magnesium ions that form deactivating soap scums.

A **detergent** is the sodium salt of a sulfonic acid. It has similar cleansing properties to soap but has many significant advantages. Calcium and magnesium ions have no effect on the cleansing properties of a detergent, since their calcium and magnesium salts are water soluble.

$$CH_3CH_2CH_2CH_2CH_2CH_2CH_2CH_2CH_2CH_2CH_2CH_2CH_2CH_2CH_2CH_2C-\overset{\overset{O}{\parallel}}{O}{}^{\ominus} \ Na^{\oplus}$$

a soap
(the sodium salt of a fatty acid)

$$CH_3CH_2CH_2CH_2CH_2CH_2CH_2CH_2CH_2CH_2CH_2CH_2-O-\underset{\underset{O}{\parallel}}{\overset{\overset{O}{\parallel}}{S}}-O^{\ominus} \ Na^{\oplus}$$

a detergent
(a sodium alkyl sulfate)

NONSAPONIFIABLE LIPIDS

Probably the most important lipid in this class, and the one you will work with in this experiment, is cholesterol. Cholesterol is a steroid. **Steroids** are compounds based on the four-ring structure shown below.

Cholesterol is synthesized by animals and used as the starting compound to make other steroids such as hormones, several vitamins, and bile salts. Cholesterol, some produced by your body and some obtained from your diet, circulates in the blood in several forms in combination with protein. Heart disease appears to be linked with high blood levels of low-density lipoproteins (LDL or "bad" cholesterol) and low levels of high-density lipoproteins (HDL or "good" cholesterol). Food fats (triglycerides) raise blood cholesterol much more than does food cholesterol. Cholesterol is found in bile liquid and is the principal component of gallstones.

steroid fused-ring structure   cholesterol

Procedure

> **CAUTION: Most organic compounds are flammable. Ether, used in Part IV, is EXTREMELY flammable. Keep away from all flames.**
>
> **Bromine and its vapors (Part V) are very irritating to the skin, eyes, and lungs. Use only in a fume hood.**

I. SOLUBILITY OF LIPIDS

1. Place 4 dry test tubes in a rack, number them, and add 1 mL of the following solvents.
   CAUTION: No flames should be present.
   - No. 1: Deionized water
   - No. 2: Ethanol
   - No. 3: Heptane

2. Add about 10 drops of cottonseed oil, which has solubility properties like those of all fats and oils, to each tube, and stir vigorously. Record the solubility (S or I) on the Report Sheet.

3. **Disposal:** Discard No. 2 (ethanol) and No. 3 (heptane) in the "Waste Organics" container.

II. IDENTIFICATION TEST FOR GLYCERIDES—ACROLEIN TEST

> **CAUTION: Do this part under a fume hood as acrolein is toxic.**

1. Place about 2 or 3 drops of cottonseed oil in a small beaker and add a few crystals of potassium hydrogen sulfate, $KHSO_4$.

2. Heat gently until a distinguishing odor is noticeable. Odor can be detected by **carefully** wafting the vapor toward your nose. **CAUTION: At this point, stop heating the beaker.**

3. Repeat Steps 1 and 2, substituting glycerol for cottonseed oil.

4. **Disposal:** Add soap, scrub, and pour in sink.

III. TEST FOR PEROXIDES

1. Add about 10 drops of an old sample of a highly unsaturated oil like safflower, sunflower, or corn oil to a 125-mm test tube. Add 3 mL of a 3:2 v/v mixture of glacial acetic acid and dichloromethane, and stir until the oil dissolves.

2. Add 3 drops of saturated aqueous potassium iodide (KI) solution and, with occasional shaking, allow the mixture to stand for 1 minute. If peroxides are present, the mixture will become pale yellow or brown from oxidation of iodide ion to free iodine (I_2).

3. Add 3 mL of water, shake, and then with continuous shaking add dropwise (COUNTING the drops) 0.01 M aqueous sodium thiosulfate ($Na_2S_2O_3$) solution until the yellow color *almost* completely disappears.

4. Add a drop of starch solution (note the color change) and add *dropwise* more $Na_2S_2O_3$ solution until the color just disappears. Record the total number of drops needed.

5. **Disposal:** Discard the test mixture in the "Waste Halogenated Solvents" container.

IV. KREIS TEST FOR OXIDATIVE RANCIDITY

1. **CAUTION: Ether is extremely flammable. Perform this test in a hood.** Place about 10 drops of the same old oil sample used in Part III in a 125-mm test tube. Add 4 drops of concentrated hydrochloric acid (CAUTION) and stir vigorously.

2. Add 10 drops of phloroglucinol solution (1% in diethyl ether) and stir vigorously again.

3. After 10 minutes, observe the color. A pink color indicates that the fat is slightly oxidatively degraded; a red color indicates that the fat is definitely degraded and rancid.

4. **Disposal:** Discard the test mixture in the "Waste Organics" container.

V. BROMINE ADDITION TEST FOR UNSATURATION

> **CAUTION: Do this work in a fume hood.**

In Step 4 below you will test cholesterol, but in Steps 1 through 3 you will test a sample of commercial solid shortening *and* a sample of commercial liquid shortening like safflower, sunflower, or corn oil.

Use a different medicine dropper for measuring each of the two triglyceride samples. (Liquefy the solid shortening just before use by placing it in a hot water bath on a hotplate until it just melts; do not overheat.) Use the following procedure for each of these two samples.

1. Place 5 drops of the melted solid shortening in a 125-mm test tube and add 2 mL of dichloromethane. Swirl the tube until the triglyceride dissolves.

2. While holding the tube over a white sheet of paper, swirl the tube contents smoothly, and add dropwise (COUNTING the drops) a 5% solution of bromine in dichloromethane until a yellow color (due to unreacted bromine) persists for 30 seconds. Record the number of drops added to the sample.

3. *Repeat* the above procedure using the liquid shortening.

4. Now, use the above procedure to test **cholesterol**. Since it does not melt readily, measure out 0.25 g of solid cholesterol (note its appearance), place it in a 125-mm test tube, and dissolve it in 2 mL of dichloromethane. Then use the procedure of Step 2 above, and record the number of drops of bromine solution required here.

5. **Disposal:** Discard the test mixtures in the "Waste Halogenated Solvents" container.

VI. CLEANSING PROPERTIES OF SOAP AND DETERGENT

1. Prepare about 60 mL of soap solution from small lumps of Ivory (or similar) soap. Place 5 mL of soap solution into each of 3 numbered test tubes, and add the following reagents:
   No. 1:   5 mL of deionized water
   No. 2:   5 mL of tap water
   No. 3:   5 mL of 0.1% $CaCl_2$ solution

2. Stopper each of the test tubes and shake for about 30 seconds. Compare the foam formed in each case. Save test tube No. 3 ($CaCl_2$) for use in Step 6.

3. Repeat Steps 1 and 2 with a detergent solution instead of the soap solution.

4. Pour 5 mL of 0.1% $CaCl_2$ solution into a test tube, add about 3 g of ion-exchange resin, and stopper the test tube. Shake the test tube for about 4 minutes.

5. Filter the solution containing the resin, and collect the filtrate in a test tube.

6. Place 5 mL of soap solution (from Step 1) in the filtrate, stopper the test tube, and shake for about 30 seconds. Compare the foam formed in this case to the foam formed with the untreated $CaCl_2$ solution (Step 2—test tube No. 3).

7. **Disposal:** Decant test samples into sink, but put the ion-exchange resin in the "Used Ion-Exchange Resin" container so it can be regenerated and reused.

Prelab Questions for Experiment 15

_____ 1. An unsaturated lipid
   a. is more difficult to digest than a saturated lipid.
   b. has a lower melting point than a saturated lipid.
   c. spoils less easily than a saturated lipid.
   d. is more soluble in polar solvents than a saturated lipid.

_____ 2. The acrolein test for a lipid is
   a. based entirely on the fact that lipids can be saponified.
   b. based upon the fact that certain lipids, upon heating, form glycerol, which is easily dehydrated.
   c. applicable to nonsaponifiable lipids as well as saponifiable lipids.
   d. specific for cholesterol.

_____ 3. If a commercial lipid produced a positive Kreis test, you should
   a. return it to the store where you purchased it and demand a refund.
   b. use it regularly for cooking.
   c. use it as a salad dressing.
   d. classify it as a nonsaponifiable lipid.

_____ 4. Cholesterol
   a. is a saponifiable lipid.
   b. is produced in the body as a result of the hydrolysis of a saponifiable lipid.
   c. is produced in the body and used to make hormones.
   d. is the principal component of kidney stones.

_____ 5. If a lipid is heated with $KHSO_4$ and acrolein is not produced, the lipid is probably
   a. unsaturated.
   b. saturated.
   c. saponifiable.
   d. nonsaponifiable.

6. Why is some water "hard"?

_____

Report for Experiment 15

I. SOLUBILITY OF LIPIDS

1. Data:

Lipid	Solubility in Water (I or S)	Solubility in Ethanol (I or S)	Solubility in Heptane (I or S)
Cottonseed oil			

2. Does the solubility data indicate lipids are polar or nonpolar? Explain.

_____

_____

3. What rule is illustrated here?

_____

II. IDENTIFICATION TEST FOR GLYCERIDES—ACROLEIN TEST

1. Observations of odor:

    a. Cottonseed oil _____

    b. Glycerol _____

2. Explain any differences.

_____

_____

3. Write a chemical equation that illustrates the result of heating glycerol with the catalyst potassium hydrogen sulfate ($KHSO_4$).

III. TEST FOR PEROXIDES

1. Color on addition of KI?

2. Color on addition of starch?

3. Number of drops of $Na_2S_2O_3$ solution required to remove color?

4. What explains this last color change?

IV. KREIS TEST FOR OXIDATIVE RANCIDITY

Color observations and conclusions:

V. BROMINE TEST FOR UNSATURATION

1. Data:

Lipid	Number of Drops of Br_2 solution
Commerical Solid Shortening	
Commercial Liquid Shortening	

2. a. Which of the shortenings tested is the more unsaturated?

   _____

   b. Which may contain more trans fat? Do the results obtained here help you answer that?

   _____

3. With cholesterol, what functional group was detected? _____
   How does it happen that the name "cholesterol" doesn't indicate presence of this group?

VI. CLEANSING PROPERTIES OF SOAP AND DETERGENT

1. Data for Steps 1 through 3:

Solution Added	Amount of Foam Produced with Soap	Amount of Foam Produced with Detergent
Deionized water		
Tap water		
0.1% $CaCl_2$		

2. a. Which of the solutions tested produced the most foam with soap? _____

   b. With detergent? _____

3. a. Which of the solutions tested produced the least foam with soap? _____

   b. With detergent? _____

4. Which ions are usually present in hard water? _____

5. Data for Steps 4 through 6:

Solution Added	Amount of Foam Produced with Soap
0.1% $CaCl_2$ (data from Step 1)	
0.1% $CaCl_2$ after ion-exchange treatment (Step 6)	

6. a. Did the ion-exchange treatment of the calcium chloride solution affect the amount of foam produced by the soap?

   _____

   b. Explain your result.

   _____

Related Questions for Experiment 15

1. a. Which of the solvents in this experiment is an effective grease remover?

   _____

   b. Explain why it is so effective.

   _____

   _____

2. The odor of acrolein is frequently strong in restaurants that feature charcoal-broiled food. Explain the presence of this odor.

   _____

   _____

3. How does a home water softener produce soft water?

   _____

   _____

4. During World War II, homemakers were urged to turn in bacon grease for use in the war effort. Explain how the grease could be converted into useful products.

   _____

   _____

5. Some soap manufacturers advertise their product as not leaving a soap film. What does this indicate about the structure of their product?

   _____

   _____

6. Formation of the first hydroperoxide molecule in a liquid shortening is a rare occurence, but it can trigger (via "free radicals") formation of two new hydroperoxide molecules. What does this tell you about shortening that is just slightly rancid?

7. Classify the following fatty acids as saturated, monounsaturated, or polyunsaturated (S, M, or P):

_____ a. 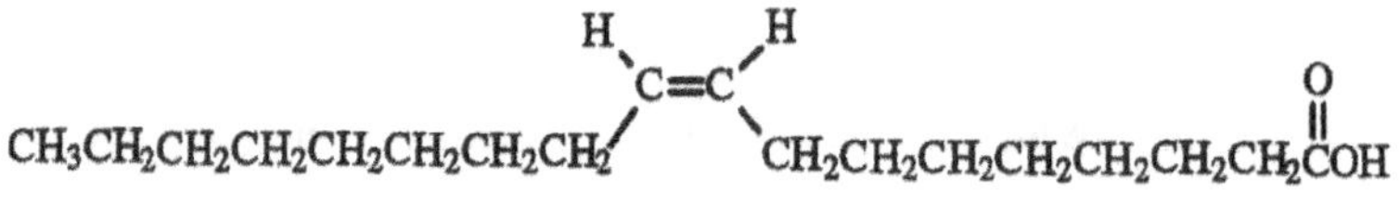 $CH_3CH_2CH_2CH_2CH_2CH_2CH_2CH_2$ $\overset{H}{\underset{}{C}}=\overset{H}{\underset{}{C}}$ $CH_2CH_2CH_2CH_2CH_2CH_2CH_2\overset{O}{\overset{\|}{C}}OH$

_____ b. $CH_3(CH_2)_{14}\overset{O}{\overset{\|}{C}}OH$

_____ c. 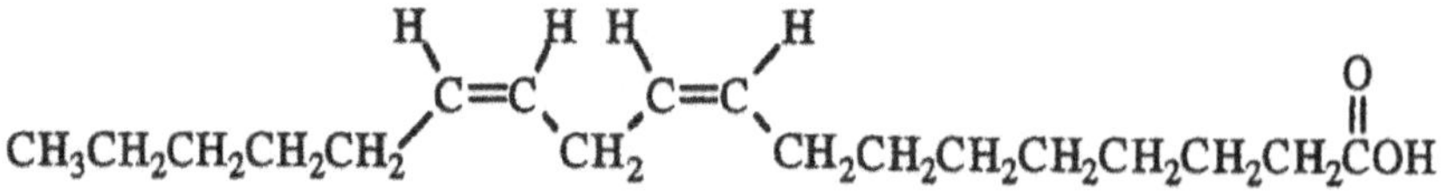 $CH_3CH_2CH_2CH_2CH_2$ $\overset{H}{\underset{}{C}}=\overset{H}{\underset{}{C}}$ CH_2 $\overset{H}{\underset{}{C}}=\overset{H}{\underset{}{C}}$ $CH_2CH_2CH_2CH_2CH_2CH_2CH_2\overset{O}{\overset{\|}{C}}OH$

8. Classify the fatty acids in Question 7 as a *cis-* or *trans-* isomer.

_____

9. Complete the following equation representing the hydrolysis of a triglyceride.

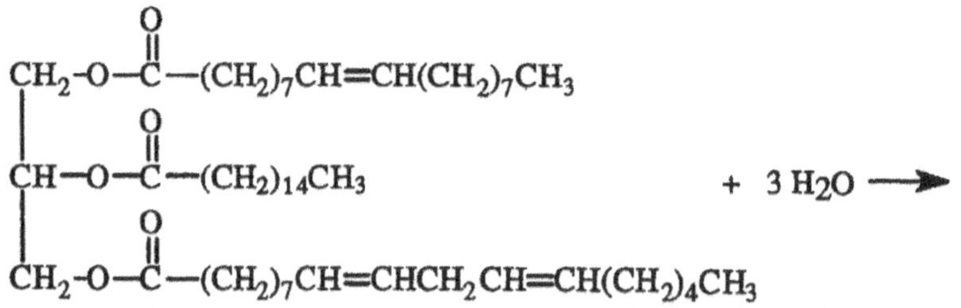

CH_2-O-$\overset{O}{\overset{\|}{C}}$-$(CH_2)_7CH=CH(CH_2)_7CH_3$

CH-O-$\overset{O}{\overset{\|}{C}}$-$(CH_2)_{14}CH_3$ $+ \ 3 \ H_2O \longrightarrow$

CH_2-O-$\overset{O}{\overset{\|}{C}}$-$(CH_2)_7CH=CHCH_2CH=CH(CH_2)_4CH_3$

10. Complete the following equation representing the complete hydrogenation of a triglyceride.

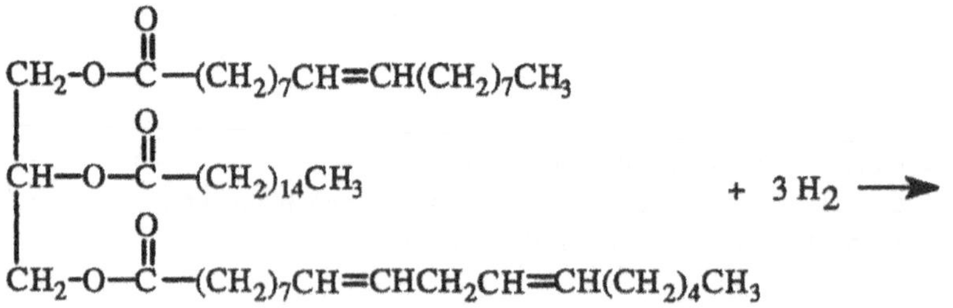

CH_2-O-$\overset{O}{\overset{\|}{C}}$-$(CH_2)_7CH=CH(CH_2)_7CH_3$

CH-O-$\overset{O}{\overset{\|}{C}}$-$(CH_2)_{14}CH_3$ $+ \ 3 \ H_2 \longrightarrow$

CH_2-O-$\overset{O}{\overset{\|}{C}}$-$(CH_2)_7CH=CHCH_2CH=CH(CH_2)_4CH_3$

Experiment 16
Proteins

Reference: *General, Organic, and Biological Chemistry: An Integrated Approach, 4th ed., Chapters 8 and 12.*

Purpose: You will perform protein identification tests, detect amino acids in orange juice, and investigate acid-base properties of proteins. You will identify factors that cause protein denaturation and determine the isoelectric point of casein, a typical protein.

Introduction

AMINO ACIDS

Proteins, essential components of the living cell, are high molecular weight compounds which, upon hydrolysis, produce amino acids. In a protein the amino acids are linked together by a series of peptide bonds. A **peptide bond** is an amide linkage; it is the functional group in proteins that can be **hydrolyzed** (split apart by water) under hot, strongly acidic conditions.

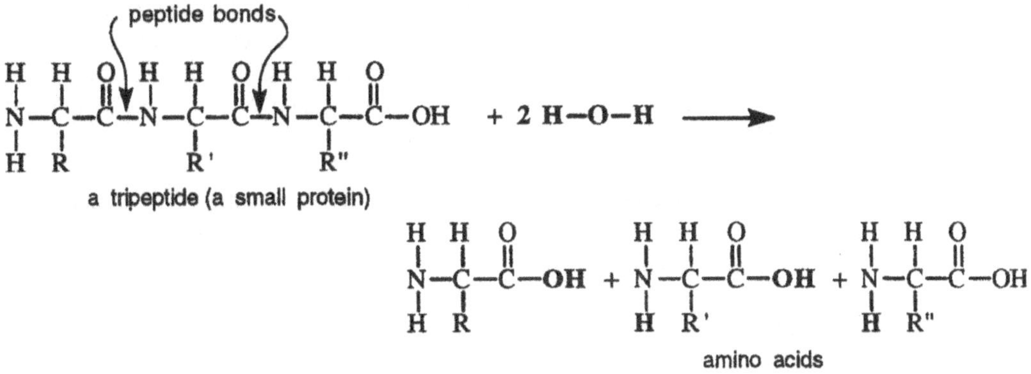
peptide bonds
+ 2 H—O—H
a tripeptide (a small protein)
amino acids

Note that, for convenience, the amino end is always depicted on the left, the carboxyl on the right.

The group represented by R may be aliphatic, aromatic, or heterocyclic and may contain additional functional groups. For example:

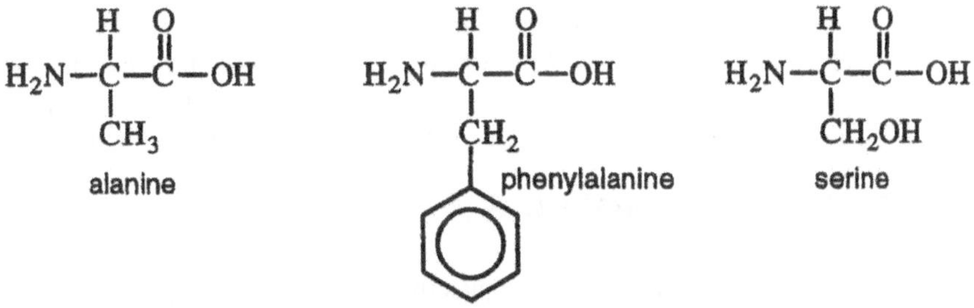
alanine
phenylalanine
serine

Human beings cannot synthesize ten of the twenty common amino acids found in proteins; these **essential amino acids** must be supplied in the diet.

The presence of both the basic amino group (–NH$_2$) and the acidic carboxyl group (–COOH) accounts for the fact that amino acids (or proteins) in water solution can act as either acids or bases. Substances that can act as either acids or bases are **amphoteric**. In this experiment you will use indicators (phenolphthalein and methyl orange) to illustrate the amphoteric properties of casein, the principle protein found in cow's milk. Methyl orange appears red in a solution having a pH less than 3, and yellow in a solution of pH greater than 4. The indicator phenolphthalein (formerly used as the active ingredient in Ex-Lax®) is colorless in a solution of pH less than approximately 8 and fuchsia in a solution having a pH greater than about 10.

THE ISOELECTRIC POINT OF A PROTEIN

Amino acids (and proteins) contain basic amino groups; when an amino group accepts a proton in an acid-base reaction, the compound becomes positively charged. When the carboxylic acid group of an amino acid (or protein) donates a proton, the substance becomes negatively charged. At a specific pH, the compound will be neutral; this pH is the **isoelectric point** of the protein (or amino acid). Each protein has a characteristic isoelectric point. At this pH, protein molecules coagulate and are removed from solution.

PROTEIN TESTS

There are several color tests that are used to detect proteins. These tests depend on the presence of a particular structure within the protein. A positive **biuret test**, for example, is given by tripeptides and all longer polypeptides. When treated with copper(II) sulfate (CuSO$_4$) in basic solution, they produce a violet color.

A second test, the **ninhydrin test**, is used extensively in chromatography to detect the presence of most amino acids. In practice, the completed chromatogram is allowed to dry, sprayed with the ninhydrin reagent, and then placed in an oven at 100°C for approximately 5 minutes to develop a violet color, caused by the presence of the amino group. Note: Proline produces a yellow color with the ninhydrin reagent.

The xanthoproteic test also detects proteins but is more often run accidentally than intentionally! When concentrated nitric acid gets on a protein, it causes a bright yellow stain, often seen (painfully) on chemists' fingers.

PROTEIN STRUCTURE

The **primary structure** of a protein refers to the exact order in which the amino acids are linked together by means of peptide bonds. The three-dimensional configuration of segments of the protein chain is called its **secondary structure**; common secondary structures are the α-helix and the β-pleated sheet. The secondary structure is formed when amino acids hydrogen bond to other amino acids farther along the polypeptide chain. The **tertiary structure** is the three-dimensional shape of the entire polypeptide chain. Globular proteins, for example, are very tightly folded into a compact spherical form. This folding results from interactions between the R side chains of amino acids and may involve hydrogen bonding as well as disulfide bondings, salt bridges, and hydrophobic interactions. Only proteins containing more than one chain, such as hemoglobin, have a **quaternary structure**; the manner in which the several peptide chains fit together describes a protein's quaternary structure.

DENATURATION OF PROTEINS

If the secondary, tertiary, or quaternary structures of a protein are disrupted, the protein loses its biological activity and is said to be **denatured**. Denaturation may result in coagulation with the protein being precipitated from solution. The factors that may cause the denaturation of proteins are pH, heat, certain organic solvents, heavy metal ions, alkaloidal reagents, reducing agents, and non-ionizing radiation. In this experiment you will study the denaturation of albumin, the protein found in egg white.

Procedure

> CAUTION: Handle hydrochloric, nitric, and sulfuric acids with care. Avoid skin contact. Wash your hands after use.
>
> Most organic compounds, such as ethanol, are flammable.

I. DETECTION OF AMINO ACIDS IN ORANGE JUICE

CAUTION: Ninhydrin may be irritating to skin and mucous membranes.

1. First, set up a developing chamber so it can equilibrate during preparation of the TLC plate. Place the provided eluent (a mixture of 95% ethanol, concentrated aqueous ammonia, and water, 8/1/1 by volume) in a 400-mL beaker, but only to a depth of about 0.5 cm so that the liquid level will not be high enough to dissolve the sample off the TLC plate. Place an aluminum foil cover on the beaker so that the vapor space and eluent can equilibrate.

2. Pour 2 mL of orange juice into a test tube and add 4 mL of ethanol to precipitate the protein.

3. Filter the solution. Only a few drops of the filtrate will be needed.

4. As in earlier TLC work (see Experiment 4), use pencil only (pen ink will smear in the eluent) to mark your thin layer chromatography (TLC) plates and do not touch the TLC plate with your fingers except at the top edge.

   To see if the amino acids in orange juice can be identified, we will use a TLC plate that is 5 cm wide and 10 cm tall, so known reference amino acids can be run as well as the orange juice filtrate prepared above. On the plate, starting 1 cm from the left (long) edge, make four very light pencil marks 1 cm apart and 1 cm from the bottom. With a micropipet, apply 2-3 μL of the orange juice filtrate at the second spot from the left, using light, quick contact with the plate so as to avoid getting too large a spot diameter. Using other, labeled micropipets, apply about the same volume of the *provided* glycine reference solution at the first spot on the left, arginine reference solution at the third spot, and asparagine reference solution at the fourth. In the Report section, begin the drawing of the TLC plate that is to be completed later and *record these spot identities and locations on it*. These reference amino acids are used because they are among the most abundant ones in orange juice.

   When the spots look dry (i.e., when the samples' solvent has evaporated), place the plate in the foil-covered beaker and wait for the eluent to reach nearly to the top. *As soon as you take the plate out*, mark the eluent front before evaporation occurs.

   Wait for the eluent on the TLC plate to dry and then, using light, quick brush strokes, brush on ninhydrin solution. Then place the plate (ON A WATCH GLASS) in an approximately 105°C oven for about 5 minutes. Circle the spots that appear, measure and record in the Report section the distance from the origin to the middle of each spot (to the nearest millimeter), and calculate the R_f value of each spot. The A-D spaces in the Report section are for the data on the amino acids found in the orange juice extract; *designate them A though D in order of increasing distance traveled.*

5. **Give your conclusions on whether or not your reference amino acids account for any of the spots from amino acids in the orange juice.**

6. In the Report section, complete the drawing of the developed TLC plate that shows its approximate spot locations (label them) and then dispose of it as your instructor specifies.

Disposal: In experiments below, dispose of test samples in a sink *unless a specific method is given.*

II. ACID-BASE PROPERTIES OF PROTEINS

1. Place 5 mL of deionized water in a test tube. Add a drop of 0.1% sodium hydroxide solution and 2 drops of phenolphthalein solution. Note the color.

2. Add about 0.1g of casein; then stir the mixture periodically for a few minutes until there is a color change. Explain your results.

3. Pour 5 mL of deionized water into each of two test tubes; then add a drop of 0.1% hydrochloric acid and 5 drops of methyl orange solution to each. Note the color.

4. To one of the tubes add about 0.1g of casein; then stir the mixture periodically for about 10 minutes until there is a color change. Compare the color produced to the color of the other test tube from Step 3.

III. THE ISOELECTRIC POINT

1. Pour about 50 mL of skim milk or an aqueous solution of powdered milk into a 150-mL beaker. Record the pH of the milk. *For best results use a pH meter*, but indicator paper having approximately a 4.5 to 8.5 range will provide satisfactory readings.

2. Place a stirring bar in the beaker and place the beaker on a magnetic stirrer. Adjust the stirrer so that the bar is rotating at moderate speed. Use a stirring rod if a magnetic stirrer is not available.

3. Add 0.1 M acetic acid drop-by-drop from a medicine dropper until the proteins (casein) begin to coagulate. The coagulation point can be detected by swirling the beaker and observing whether particles adhere to its side. Record the pH of the milk at that point; it is the isoelectric point.

IV. DENATURATION OF PROTEINS

Disposal: In Parts A and D, pour final test mixtures in "Waste Inorganics" container.

A. pH

1. Pour about 3 mL of 2% albumin solution into each of three test tubes in a test-tube rack, number the tubes and add the following reagents: (**CAUTION!**) Wash your hands.
   No. 1:   1 mL of 3M HCl
   No. 2:   1 mL of 3M HNO_3
   No. 3:   1 mL of 3M H_2SO_4

2. Stir each test tube. Record your observations. Wash your hands.

B. Heat

1. Pour about 5 mL of albumin solution into one labeled test tube and 5 mL of skim milk (or powdered milk solution) into another. Heat them in a boiling water bath over a hot plate for about 10 minutes.

2. Carefully note whether or not any coagulants appeared in the tubes.

C. Organic Solvents

1. Pour about 5 mL of albumin solution into a test tube and, with stirring, add 5 mL of 95% ethanol.

2. Compare the appearance with that of the untreated albumin solution.

D. Heavy Metal Ions

1. Pour 3 mL of albumin solution into each of two test tubes and place them in a rack. Number the test tubes, and add 10 drops of the following reagents:

   No. 1: 1% silver nitrate ($AgNO_3$) solution. (**CAUTION: It will stain clothing and/or skin after exposure to light. It is what happens in photographic film.**)
   No. 2: 1% sodium chloride (NaCl) solution

2. Stir each test tube. Record your observations.

E. Alkaloidal Reagents

1. Pour about 5 mL of albumin solution into a test tube and add about 1 mL of a 10% tannic acid solution.

2. Stir the test tube. Record your observations.

V. TESTS FOR PROTEINS

A. Biuret Test

1. Pour 3 mL of albumin solution into a test tube and add an equal volume of 10% NaOH solution with stirring. **CAUTION: 10% NaOH is caustic. Wear protective gloves.**

2. Add 5 drops of 0.5% $CuSO_4$ solution. Allow several minutes for the reaction to occur. Then assess and record the color.

3. **Disposal:** Pour contents of the test tube in "Waste Inorganics" container.

B. Ninhydrin Test

See its use in Part I.

C. Xanthoproteic Reaction

CAUTION: Handle concentrated nitric acid with care. Wash your hands after use.

1. Add 0.1 g of casein to a test tube and add 2 drops of concentrated nitric acid. Note the color that develops.

2. Add 15 drops of 10% sodium hydroxide (NaOH) and stir.

3. Record your observations on color changes.

Prelab Questions for Experiment 16

True or False:

_____ 1. All amino acids will produce a positive biuret test.

_____ 2. Most amino acids will produce a positive ninhydrin test.

Multiple choice:

_____ 3. At the isoelectric point of a protein,
   a. the pH of the solution will be 7.
   b. the number of positive charges on the protein will equal the number of negative charges.
   c. the pH of the solution will always be greater than 7.
   d. the pH of the solution will always be less than 7.

_____ 4. Ethanol is added to the orange juice in Part I of this experiment to
   a. precipitate the protein.
   b. detect the amino acids.
   c. develop the chromatogram.
   d. emulsify the orange juice.

5. A chromatogram is performed on a mixture of amino acids. Given the following data, calculate the R_f value for each amino acid.

		R_f Value
Eluent distance	4.5 cm	
Amino acid "A" distance	0.5 cm	_____
Amino acid "B" distance	1.2 cm	_____
Amino acid "C" distance	3.6 cm	_____

Calculations:

Name ______________________ Section ____________ Date ______________

Report for Experiment 16

I. DETECTION OF AMINO ACIDS IN ORANGE JUICE

1. Draw your chromatogram here.

2. Label each orange juice amino acid you have circled A, B, C, etc., in order of increasing travel distance, and calculate the R_f value for each sample.

SAMPLE	*Distance (mm)*	*R_f Value*
Eluent distance	______	1.00
Amino acid "A" distance	______	______
Amino acid "B" distance	______	______
Amino acid "C" distance	______	______
Amino acid "D" distance	______	______
Reference 1 (glycine)	______	______
Reference 2 (arginine)	______	______
Reference 3 (asparagine)	______	______

3. Does the R_f of any reference amino acid match that of any orange juice amino acid? If so, which?

II. ACID-BASE PROPERTIES OF PROTEINS

1. Data:

   a. Color of phenolphthalein/NaOH solution

   —before adding casein ______________

   —after adding casein ______________

b. Color of methyl-orange/HCl solution

—before adding casein   _____

—after adding casein   _____

2. Why did the solutions change color?

_____

3. Which property of proteins is illustrated by the above?

_____

4. Why was methyl orange used in place of phenolphthalein in the second part of this procedure?

_____

III. THE ISOELECTRIC POINT OF CASEIN

1. Data:

Initial pH of milk   _____

Final pH of milk   _____

2. What is the isoelectric point of casein?   _____

IV. DENATURATION OF PROTEINS

A. pH

Observations

1. Tube No. 1: 3 M HCl

_____

_____

2. Tube No. 2: 3 M HNO_3

_____

_____

3. Tube No. 3: 3 M H_2SO_4

_____

_____

B. Heat

Observations of heated solutions:

1. Albumin solution

   _____

   _____

2. Skim milk

   _____

   _____

C. Organic Solvents

Observations of ethanol added to albumin

_____

D. Heavy Metal Ions

1. Observations:

   a. Tube No. 1: $AgNO_3$ solution

   _____

   b. Tube No. 2: NaCl solution

   _____

2. Explain the differences in your results.

   _____

   _____

E. Alkaloidal Reagents

Observations of tannic acid added to albumin.

_____

V. TEST FOR PROTEINS

A. *Biuret Test*

Observations

_____

_____

_____

_____

B. *Ninhydrin Test*

Observations based on work in Part I.

_____

_____

_____

_____

C. *Xanthoproteic Reaction*

Observations

Name ______________________ Section __________ Date __________

Related Questions for Experiment 16

1. Explain why precipitation takes place at the isoelectric point of an amino acid or protein.

2. Why are lead compounds poisonous?

3. a. Explain why dilute silver nitrate is applied to the eyes of newborn infants.

   b. Explain why dilute, rather than concentrated, silver nitrate must be used when applied to the eyes.

4. Why is the isopropyl alcohol that is used in rubbing alcohol a disinfectant?

5. Why are tannic acid and picric acid used in commercial products for the treatment of burns?

6. Based on your laboratory observations, how might ninhydrin be used in forensic chemistry to solve crimes?

7. a. NutraSweet®, the synthetic sweetener aspartame, is a dipeptide (protein) that is used in low-calorie beverages. Explain why beverages containing NutraSweet® lose their sweetness after several months of storage.

_____

_____

b. Explain why NutraSweet® loses its sweetness when heated and cannot be used in cooking.

_____

_____

Experiment 17

Enzymes

Reference: General, Organic, and Biological Chemistry: An Integrated Approach, 4th ed., Chapter 12.

Purpose: You will perform tests to determine the chemical nature of enzymes and the specificity of enzyme action. You will investigate the effects of temperature, pH, substrate concentration, and enzyme concentration on enzyme activity.

Introduction

Enzymes are either water-soluble or fat-soluble proteins that function as biological catalyst for the reactions involved in metabolism. They may be **simple proteins** composed entirely of amino acids or may be **conjugated proteins** that require a nonprotein group for their biological activity. Enzymes are similar to inorganic catalysts in that they catalyze reactions by lowering the activation energy. However, they differ from inorganic catalysts in their specificity. For example, platinum will catalyze several different types of reactions, but a given enzyme will catalyze only one type of reaction. This **specificity** of enzymes is explained by the **lock-and-key theory**. This theory assumes that there is a specific area on the surface of the enzyme, called the **active site**, to which the substance acted upon (the **substrate**) attaches during a reaction. The configuration (shape) of the enzyme's active site and of the substrate must be complementary in much the same way as a key fits a lock. The configuration of the lock is specific for only one key; no other keys will turn the lock. In summary, enzymes are proteins whose catalytic activity is dependent upon their specific three-dimensional shape.

Factors such as pH, temperature, solvents, and salt concentrations can alter the secondary, tertiary, and quaternary structure of a protein. Such factors, therefore, will have an effect on the activity levels of enzymes. The **optimum pH** of an enzyme is that pH at which it will function most effectively, and the **optimum temperature** is the temperature of most efficient catalytic activity.

The three enzymes used in this experiment—urease, pepsin, and amylase—are representative of all enzymes; they were selected because of their availability and the ease with which their activity may be observed in simple laboratory experiments.

Urease catalyzes a very specific chemical reaction: the hydrolysis of urea.

$$\text{H}_2\text{N}-\overset{\overset{\text{O}}{\|}}{\text{C}}-\text{NH}_2 \;+\; \text{H}_2\text{O} \;\xrightarrow{\text{urease}}\; 2\,\text{NH}_3\,(g) \;+\; \text{CO}_2\,(g)$$

One of the products of this reaction, ammonia, increases the pH of the urease solution enough so that the reaction may be followed with indicators such as phenolphthalein, which is colorless in acid solution and pink to red in base solution.

Pepsin is found in the stomach and urine; it hydrolyzes peptide bonds between two aromatic amino acids or between glutamic acid and an aromatic amino acid. Its optimum pH varies from 2.0 to 4.5.

Amylase, found in the urine and saliva, hydrolyzes starch (a polysaccharide) to form maltose (a disaccharide). The digestion (hydrolysis) of starch may be followed by periodically testing the solution with an iodine solution, which produces an identifying color as illustrated in the following word equation.

$$\text{starch} \xrightarrow{\text{amylase}} \text{amylodextrin} \longrightarrow \text{erythrodextrin} \longrightarrow \text{achrodextrin} \longrightarrow \text{maltose}$$

starch (blue) → amylodextrin (blue) → erythrodextrin (red) → achrodextrin (colorless) → maltose (colorless)

(The "colorless" final stage will likely appear slightly yellowish from the iodine test reagent.)

Because the activity of amylase is increased in the presence of chloride ion, urine would be mixed with a sodium chloride solution before studying its properties. The "urine solution" used here is artificial (see below) and is premixed with sodium chloride. Amylase is also activated by calcium ion found in the urine.

Procedure

The "urine solution" used in this experiment is a substitute mixture of harmless, pure chemicals and poses no biological hazard. Therefore you do not need to use latex gloves for any part of this experiment, but if you were using actual urine samples, you should. If actual urine were to be used, it should be mixed with an equal volume of 0.9% NaCl solution, but NaCl has been incorporated into the artificial "urine solution" used here.

Disposal: In experiments below, dispose of test samples in a sink *except for that in Part I, which should be poured into the "Waste Inorganics" container.*

I. THE CHEMICAL NATURE OF ENZYMES (BIURET TEST)

1. Pour about 3 mL of 1% urease solution into a test tube and add an equal volume of 10% NaOH solution with stirring. **CAUTION: 10% NaOH is caustic. Wear protective gloves.**

2. Add 5 drops of 0.5% $CuSO_4$ solution. Allow several minutes for the reaction to occur. Then assess and record the color. (See also Experiment 16 for discussion of the biuret test.)

II. SPECIFICITY OF ENZYME ACTION

1. Label three 16 × 150-mm test tubes and put 5 mL of 2% urea solution + 3 drops of phenolphthalein indicator into each of them. Swirl. Then add the following to *the indicated tubes*:
   No. 1: 3 mL of deionized water (control)
   No. 2: 3 mL of 1% urease solution
   No. 3: 3 mL of 1% pepsin solution

2. Swirl each test tube. Observe the color in each test tube. [Remember that phenolphthalein is colorless in neutral and acidic solutions and pink to red in basic (alkaline) solutions.]

III. THE EFFECT OF SUBSTRATE CONCENTRATION

1. Label four 25 × 150-mm test tubes 1-4 and add the following to *the indicated tubes*:
   No. 1: 20 mL of deionized water + 3 drops of phenolphthalein indicator (control)
   No. 2: 10 mL of 2% urea solution + 10 mL of deionized water + 3 drops of phenolphthalein indicator
   No. 3: 5 mL of 2% urea solution + 15 mL of deionized water + 3 drops of phenolphthalein indicator
   No. 4: 30 mL of 1% urease solution

2. Place all four test tubes in a water bath maintained at 40°C, using a hot plate, for 5-10 minutes to thermally equilibrate. Then transfer approximately 10-mL portions of the enzyme solution in **No. 4** into each of three 16 × 150-mm test tubes (just match the solution heights in the tubes) and, as **simultaneously as possible**, pour one of these equal-sized portions of the enzyme solution into each of tubes **No. 1, 2, and 3** that are in the 40°C bath. **Note the time.**

3. Constantly observe tubes **No. 2 and 3**, comparing them with control tube, **No. 1**. Record for tubes **No. 2 and 3** (a) the time required for the appearance of pink coloration and (b) for reaching the maximum intensity of pink color observed within 10 minutes. Note any change in the control sample, **No. 1**.

4. Just after this 10-minute period, use a pH meter to measure the pH of the contents of tubes **No. 2 and 3**. Record your observations on the Report Sheet.

IV. TEMPERATURE AND ENZYME ACTIVITY

1. Pour 20 mL of freshly prepared 0.075% starch solution into each of three numbered 50-mL Erlenmeyer flasks.

2. Place flask No. 1 in a water bath, maintained at 55–60°C, using a hot plate. Place flask No. 2 in a water bath, maintained at 40°C, using a hot plate. Place flask No. 3 in an ice-water bath.

3. Allow 5 minutes for each of the solutions to reach the temperature of the water bath.

4. Add 15 mL of urine solution to each of the three flasks. **Mix well by swirling the flasks.** Note the time at which the urine was added.

5. After 5 minutes, test for the presence of starch in each of the solutions as follows. Place three drops of freshly prepared 0.025% iodine solution in a small (10 × 75-mm) test tube. Using a medicine dropper, add 3 drops of urine-starch solution. Use separate test tubes and a different medicine dropper for each sample. Compare the colors of the three solutions.

6. After testing the samples, return each Erlenmeyer flask to its respective water bath. Repeat the tests at 5 minute intervals until one of the solutions produces the red color characteristic of erythrodextrin. Record the time for a change to take place.

V. pH AND ENZYME ACTIVITY

It would be best to work with a lab partner in Parts V and VI of this experiment.

1. Use three 15 × 150-mm test tubes labeled 1, 2, and 3. Into each one, put 6 mL of 0.075% starch solution.

2. Add a 3-mL portion of the indicated buffer as follows:
   No. 1: pH 6 buffer
   No. 2: pH 7 buffer
   No. 3: pH 8 buffer
   Then place all three tubes in the 40°C water bath.

3. While the tubes are adjusting to the 40°C temperature, measure out three 2-mL portions of the "urine solution" in 12 × 75-mm test tubes. Add one such portion to each of the tubes **as nearly simultaneously as possible** (record the time) and then keep them in the 40°C water bath throughout the balance of this Part V work.

4. **After just 1 minute**, test for the presence of starch and its cleavage products as done in Part IV: Place three drops of freshly prepared 0.025% iodine solution in a small (12 × 75 mm) test tube. Then, using a Pasteur pipet, add 3 drops of the solution in tube No. 1 (or, successively, No. 2 and No. 3). Use separate test tubes and Pasteur pipets for each test. **Record the time when each test is conducted and when each of the below colors appear.**
   Blue, which indicates that starch is present
   Red (pink), which indicates that erythrodextrin is present
   Yellow, which indicates that cleavage of the starch has proceeded to the achrodextrin or later stage(s). (The yellow color is from the diluted iodine reagent.)

5. After each additional **1- to 2-minute interval**, repeat the testing of the solutions in tubes No. 1 through 3 until they give indication of cleavage through to the achrodextrin or later stage. You do not need to repeat the testing beyond the time when the first two tube contents test red or yellow.

VI. THE EFFECT OF ENZYME CONCENTRATION

1. The "urine solution" is the one that was used in Parts IV and V. Here it will be used as is and also in diluted form, which should be prepared by adding 9 mL of deionized water to 1 mL of the stock solution.

2. Label two 16 × 150-mm test tubes and into *both* put 6 mL of 0.075% starch solution and 3 mL of pH 7 buffer. Then put in, **as nearly simultaneously as possible** (record the time), to:
   No. 1: 2 mL of the **diluted** "urine solution."
   No. 2: 2 mL of the **undiluted** "urine solution."
   Then place both tubes in the 40°C water bath.

3. Using the procedure of Parts IV and V above, test for starch degradation **after 1 minute.** Repeat the tests at 1- to 2-minute intervals until one of the two solutions produces the erythrodextrin red color. **Record the time when the red color first appears.**

Prelab Questions for Experiment 17

1. Starch is being hydrolyzed in the presence of amylase to form erythrodextrin. How would you test for the presence of erythrodextrin?

   _____

   _____

2. When urea is hydrolyzed, the pH of the solution rises. Why?

   _____

   _____

3. In Part III, The Effect of Substrate Concentration, how can you determine whether solution 2 or 3 reacts faster?

   _____

   _____

4. What other body fluid could have been substituted for the urine solution in Parts IV through VI of this experiment and still have yielded the same results?

   _____

   _____

5. Why is the biuret test performed on the urease solution in Part I?

   _____

Report for Experiment 17

I. THE CHEMICAL NATURE OF ENZYMES (Biuret Test)

1. Observations

_____

_____

_____

2. Based on your observations, to what class of compounds do enzymes belong?

_____

II. SPECIFICITY OF ENZYME ACTION

1. Observations:

   a. No. 1—water (control)

   _____

   b. No. 2—urease + urea

   _____

   c. No. 3—pepsin + urea

   _____

2. a. Which of the two enzymes catalyzed a chemical reaction?

   _____

   b. How do you know a reaction took place?

   _____

   c. What were the products of the reaction?

   _____

3. What was the purpose of the control test tube (No. 1—water)?

   _____

III. THE EFFECT OF SUBSTRATE CONCENTRATION

    1. Observations:

        a. No. 1—water (control)   pH _____

           _____

        b. No. 2—10 mL of 2% urea + urease   pH _____

           _____

        c. No. 3—5 mL of 2% urea + urease   pH _____

           _____

    2. a. Which of the above reactions proceeded at the faster rate?

           _____

        b. Explain, with the use of an equation, how the pH enabled you to arrive at your conclusion.

           _____

           _____

           _____

IV. TEMPERATURE AND ENZYME ACTIVITY

    1. Data:

Solution	Temperature	Time urine was added	Time iodine solution is red	Time iodine solution is colorless
No. 1				
No. 2				
No. 3				

    2. Which reaction temperature hydrolyzed starch the fastest? _____

    3. Explain in terms of molecular structure why the other two solutions reacted more slowly.

        _____

        _____

V. pH AND ENZYME CONCENTRATION

1. Data:

Solution	pH	Time urine was added	Time iodine solution is red	Time iodine solution is colorless
No. 1	6			
No. 2	7			
No. 3	8			

2. In which of the three solutions was the starch hydrolyzed most rapidly? _____

3. Why is enzyme function affected by pH?

_____

_____

VI. THE EFFECT OF ENZYME CONCENTRATION

1. Data:

Solution	Time urine was added	Time iodine solution is red	Time iodine solution is colorless
Diluted urine solution			
Urine solution			

2. Which solution hydrolyzed starch faster?

_____

3. Explain your results.

_____

_____

Name ______________________ Section __________ Date __________

Related Questions for Experiment 17

1. Of experiments like those done in Parts II through VI, which provide support for the lock-and-key theory of enzyme action?

2. Why will lead nitrate solution slow or stop the hydrolysis of starch?

3. Why will the addition of oxalate ion to the urine slow or stop the hydrolysis of starch? (Hint: see the study of oxalate ion in Experiment 11.)

Experiment 18

Urine

Purpose: You will investigate the properties of a commercial urine sample with respect to pH, specific gravity, and some inorganic and organic compounds. You will diagnose a simulated "pathological" urine specimen for the above properties.

Introduction

Urine is an aqueous solution of various organic and inorganic substances that are the waste products of metabolism or products from ingested foods and fluids. The adult human body excretes an average of 800 to 2000 milliliters of urine in 24 hours. More urine is formed during the waking hours than during an equivalent period of sleep. The total quantity of waste products excreted during the two periods is about the same, but the night urine is more concentrated. The average composition of urine of a healthy adult can be approximated as follows:

Water	95%	Sodium, Na^+	0.35%
Urea	2%	Potassium, K^+	0.15%
Uric acid	0.05%	Ammonium, NH_4^+	0.04%
Creatinine	0.075%	Calcium, Ca^{2+}	0.015%
Glucose	0%	Magnesium, Mg^{2+}	0.006%
Protein	0%	Chloride, Cl^-	0.60%
Ketone bodies	0%	Sulfate, SO_4^{2-}	0.18%
		Inorganic phosphorus, PO_4^{3-}, HPO_4^{2-}, $H_2PO_4^-$	0.10%

Urea is the major organic component in urine. Excretion of urea is the principal method for removing nitrogen from the body.

The clinical purpose for analyzing urine is to determine the existence of a body malfunction. The presence of unusually large or small quantities of normal metabolic end products or the existence of abnormal substances in the urine may indicate a metabolic disorder. The disturbance may be a malfunction directly associated with the kidney, or it may be a condition in which the kidneys function normally but excrete a specific metabolic product associated with a particular disease.

For example, **phenylketonuria**, or **PKU**, is a condition first indicated by the presence of phenylpyruvic acid in the urine. The continuous presence of glucose in the urine is an indication of **diabetes mellitus**. In addition the presence of acetone and other **ketone bodies** reveals faulty carbohydrate metabolism, a condition associated with **diabetes** and starvation. Huge amounts of the protein albumin in the urine may be an indication of renal (kidney) disorder.

The physical properties of urine may also provide information regarding a body malfunction. Normal freshly voided urine has an aromatic odor believed to be caused by the presence of rapidly evaporating acids. The urine of persons with bladder infections is usually foul-smelling because of the presence of bacteria. Persons with diabetes mellitus may have urine that has a fruity odor due to the presence of acetone.

The color of urine depends upon the concentration of its components, but it is usually amber because of the presence of a yellow pigment, urochrome. Excessive amounts of bile pigments will impart a dark yellow (almost orange) color to the urine, an indication of **obstructive jaundice**. The appearance of blood in the urine may be caused by kidney damage or by an infection of the urinary tract.

The normal specific gravity of urine is between 1.005 and 1.030. A continuously low reading may indicate **diabetes insipidus**, a disease caused by the impairment of the normal functioning of an antidiuretic hormone, whose function is to decrease the flow of urine in the kidney. Low specific gravity readings may also indicate that the kidney has lost its ability to concentrate the urine because of tubular damage.

High specific gravity readings occur in patients with diabetes mellitus, adrenal inefficiency, hepatic disease, or congestive cardiac failure. The specific gravity of urine is elevated because of the excessive loss of water caused by sweating, fever, vomiting, or diarrhea. Such extensive loss of water results in a high concentration of urinary constituents, particularly glucose and protein, and hence a high specific gravity.

The pH of normal urine can vary from 4.5 to 8.0. Persons on a normal diet excrete urine with a pH of approximately 6.0. Excessively acid urine, with a pH lower than 6.0, may be excreted by persons on a high protein diet. In addition, patients having acidosis or unchecked diabetes mellitus may excrete urine containing large amounts of acid. Normal urine is usually alkaline after meals, and individuals who are vegetarians usually void alkaline urine. Consistently high alkaline urine is an indication of a urinary tract infection.

Some substances not completely metabolized by the body are excreted unchanged in the urine. Quinine, nicotine, and excess vitamin C are typical examples.

Some of the methods you will employ in detecting the presence of inorganic ions (normal components of urine), glucose, and acetone are identical to those you have performed in previous experiments. The Keto-Diastix Reagent Strips will be used to detect both glucose and ketones. The Keto-Diastik Reagent Strips can be found through the following web sites: www.drugstore.com and www.diabeticdrugstore.com.* The strip is impregnated with the enzyme glucose oxidase to detect glucose and with nitroprusside for the detection of ketone bodies.

---

* Wiley and the authors have made every effort to ensure that these links are actively maintained. Occasionally, some sites may be taken offline by their owners without our knowledge.

Procedure

> **CAUTION: We require that disposable latex or vinyl gloves be worn during this experiment. Remember: Wearing gloves protects you, but to protect others your gloves must be washed after each use.**

Use both commercial urine and a synthetic pathological sample supplied by your instructor in performing Parts I through IV.

I. pH

Use extended-range pH paper to check the pH of the urine sample. Record the pH. Note the color, odor, and cloudiness of the urine. Dispose of the paper in a waste basket.

II. SPECIFIC GRAVITY

Use a urinometer to measure the specific gravity of the urine specimen. See Experiment 1, Part VI, for the proper procedure. Record the specific gravity. Save the urine specimen and use it in the below tests.

III. INORGANIC COMPONENTS OF URINE

CAUTION: Nitric acid is caustic. Wash your gloved and bare hands.

A. Chloride

1. Pour 2 mL of urine into a test tube and add 1 mL of 6 M nitric acid. **CAUTION!** Wash your hands.

2. Add 1 mL of 0.1 M silver nitrate solution. A white precipitate indicates the presence of chloride in the urine.

3. **Disposal:** Discard the mixture into the "Waste Inorganics" container.

B. Sulfate

CAUTION: Hydrochloric acid is corrosive.

1. Pour 2 mL of urine into a test tube, and add 1 mL of 6 M HCl.

2. Add 1 mL of 0.5 M $BaCl_2$. A white precipitate indicates the presence of sulfate.

3. **Disposal:** Discard the mixture into the "Waste Inorganics" container. Wash your hands.

C. Phosphate

CAUTION: This part of the experiment should be performed in a fume hood. Concentrated nitric acid is corrosive. Wash your gloved and bare hands.

1. Pour 3 mL of urine into a test tube; then add about 3 drops of concentrated nitric acid (**CAUTION**) and 2 drops of 2.5% ammonium molybdate reagent. Wash your hands.

2. Place the test tube in a hot-water bath using a hot plate. The formation of a yellow-green precipitate (ammonium phosphomolybdate) indicates the presence of phosphate ion.

3. **Disposal:** Pour the contents of the test tube in the "Waste Inorganics" container.

D. Calcium

CAUTION: Potassium oxalate is toxic.

1. Pour 5 mL of urine into a test tube and add 5 drops of 20% potassium oxalate. A white precipitate indicates the presence of calcium ion.

2. **Disposal:** Dispose of the contents of the test tube in the "Waste Inorganics" container.

E. Ammonium Ion

CAUTION: Sodium hydroxide is caustic. Wash your gloved and bare hands.

1. Pour 5 mL of urine into a test tube and add 2 mL of 6 M sodium hydroxide. (**CAUTION!**) Wash your hands.

2. Using a hot water bath, gently heat the liquid in the test tube.

3. Place a strip of moist red litmus paper across the opening of the test tube. **Do not allow the litmus to touch the sides of the tube!**

4. Carefully smell the gas (NH_3) evolved, using the wafting technique demonstrated by the instructor.

5. **Disposal:** Discard the contents of the test tube in the "Waste Bases" container.

IV. ORGANIC COMPONENTS OF URINE

A. Albumin (a Protein)

1. Pour 5 mL of urine into a test tube and heat it in a hot water bath.

2. If the urine becomes cloudy, add several drops of 5% acetic acid and stir. If the cloudiness remains, albumin is present; if it disappears, the cloudiness was due to the presence of phosphates.

3. **Disposal:** Discard the contents of the test tube in a sink.

B. Glucose—Clinitest Reagent

CAUTION: Clinitest tablets contain sodium hydroxide. Avoid contact with skin, eyes, mucous membranes, and clothing.

1. Pour 1 mL of urine into a test tube and add 2 mL of deionized water.

2. Place the test tube in a rack and drop a Clinitest tablet into the solution. **(CAUTION!) Do not handle the test tube until 15 seconds after the reaction has stopped.** Wash your hands.

3. Remove the test tube from the rack, stir, and compare the color produced with the color chart.

4. **Disposal:** Discard the contents of the test tube in the "Waste Bases" container.

C. Ketone Bodies—Legal's Test

CAUTION: Glacial acetic acid is corrosive and concentrated ammonium hydroxide is caustic. Both have strong, irritating odors. Wash your gloved and bare hands.

1. Pour 2 mL of urine into a test tube. Then add 2 drops of glacial acetic acid and 3 drops of 5% sodium nitroprusside solution.

2. Incline the test tube and pour 1 mL of concentrated ammonium hydroxide down the side of the tube so that two layers are formed. A red color at the boundary of the layers indicates the presence of acetone bodies.

3. **Disposal:** Discard the contents of the test tube in the "Waste Bases" container.

D. Glucose and Acetone—Keto-Diastix Reagent

1. Pour 3 mL of urine into a test tube. Dip the reagent end of a Keto-Diastix strip in the specimen for **2 seconds** and remove it immediately.

2. Tap the strip against the sides of the test tube to remove the excess urine.

3. Compare the reagent side of the Keto-Diastix to the appropriate color chart. Take a ketone reading at **15 seconds** and a glucose reading after **30 seconds**. The time at which you take these readings is critical for obtaining accurate results.

4. **Disposal:** Discard the test strip in a waste basket and the contents of the test tube in a sink.

Name ______________________ Section ____________ Date ______________

Prelab Questions for Experiment 18

What physiological conditions are associated with the following:

a. Foul-smelling urine

____

____

b. Appearance of blood in the urine

____

____

c. Excessively high specific gravity readings

____

____

d. Continuously low specific gravity readings

____

____

e. Consistently alkaline urine

____

____

f. Deep yellow color in the urine

____

____

Name ______________________ Section ____________ Date ____________

Report for Experiment 18

I-III. URINE ANALYSIS—INORGANIC COMPONENTS

1. Data

Sample	*pH*	*Specific Gravity*	Cl^- (+ or –)	SO_4^{2-} (+ or –)	PO_4^{3-} (+ or –)	Ca^{2+} (+ or –)	NH_4^+ (+ or –)

2. Describe the color, odor, and clarity of the urine tested.

____

____

____

3. a. What precipitate was formed when the urine was treated with silver nitrate?

____

b. Write an ionic equation to illustrate the precipitate's formation.

____

4. a. What precipitate was formed when the urine was treated with barium chloride?

____

b. Write an ionic equation to illustrate the precipitate's formation.

____

5. a. What precipitate was formed when the urine was treated with potassium oxalate?

____

b. Write an ionic equation to illustrate the precipitate's formation.

____

6. a. Explain your observations of the red litmus paper when you were testing for the presence of NH_4^+.

_____

_____

b. What odor did you detect while performing the above test?

_____

IV. URINE ANALYSIS—ORGANIC COMPONENTS

1. Data:

Sample	Albumin (+ or −)	Glucose		Ketone Bodies	
		Clinitest (%)	Diastix (%)	Legal's (+ or −)	Diastix (%)

2. Diagnose a possible physiological condition of the patient whose "pathological" urine you analyzed.

_____

_____

_____

_____

_____

Name ______________________________ Section ____________ Date ____________

Related Questions for Experiment 18

1. Explain how a urine analysis may help identify a drug addiction.

____

____

____

2. a. Does a positive glucose test always indicate a diabetic condition?

____

   b. Explain your answer in question 2a.

____

____

____

3. What does the presence of ketone bodies indicate?

____

____

4. Give the names and structural formulas of ketone bodies that may be found in the urine.

5. What does a positive albumin test indicate?

____

____

6. a. Water-soluble vitamins in excessive doses (beyond the minimum requirements) are excreted in the urine. Will excessive doses of these vitamins likely be of benefit to an individual?

_____

b. Explain your answer in Question 6a.

_____

_____

_____

c. Discuss the possible harm that may be produced as a result of excessive intake of vitamins.

_____

_____

_____

7. a. Write the structural formula of urea.

b. To what class of compound (alcohol, amine, etc.) does urea belong?

_____

Nucleosides

Reference: General, Organic, and Biological Chemistry: An Integrated Approach, 4th ed., Chapters 10 and 13. See also Experiment 13.

Purpose: You will study the nature of the nucleosides found in an exotic plant.

Introduction

The plant to be studied is an exotic that grows only above 90° north latitude and, as you can well imagine, is harvested only at considerable risk. From the plant material collected, we have two fractions that are available for study.

Here are some structures that illustrate the terms helpful in the studies to be made:

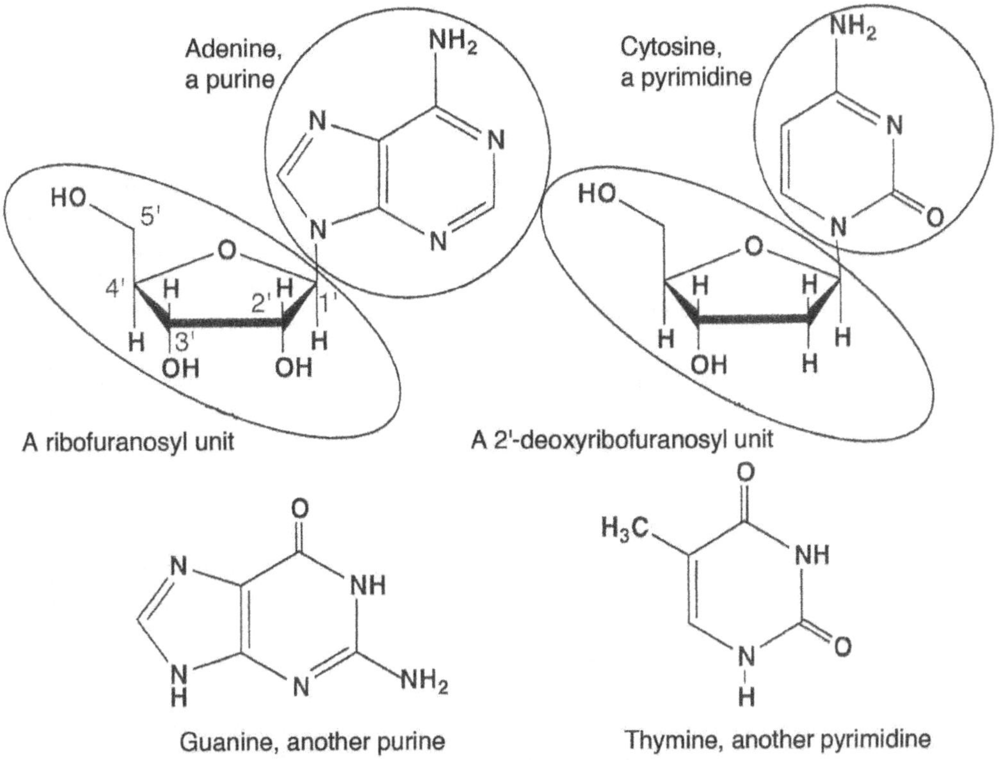

Adenine, a purine

Cytosine, a pyrimidine

A ribofuranosyl unit

A 2'-deoxyribofuranosyl unit

Guanine, another purine

Thymine, another pyrimidine

The first of the plant fractions obtained for study is ground seed material, which will be tested for the presence of purine-type deoxynucleoside units by use of the Dische reaction. It detects structures of this type whether free or incorporated into the polymer deoxyribonucleic acid (DNA) or its monomers, the deoxynucleotides (phosphate esters of deoxynucleosides). In a positive test the Dische reagent, an acidic solution of diphenylamine, turns the test mixture blue. It does not detect pyrimidine-type deoxynucleosides because they are much more resistant to hydrolytic cleavage of the bond between the sugar and the base, a hydrolysis that is required for production of the blue coloration.

The second plant sample to be studied is representative of what could be obtained by acid-catalyzed hydrolysis of ground seed material, and it will be studied by thin layer chromatography. The object will be to learn something about the sugar and the purine- and pyrimidine-type bases found in the DNA of this plant. It, being so exotic as to grow in such a very unusual location, is suspected from earlier studies to have a form of DNA that utilizes only two[*] of the four usual nitrogen bases, and we are to determine which two of adenine, cytosine, and thymine are the ones present. (The possibility of guanine being one of the two present cannot be determined by the procedure used.)

Start this chromatographic study first so that the Dische test on the ground seed material can be done while waiting for the TLC plate to develop.

Procedure

I. CHROMATOGRAPHY

1. Start by setting up a development chamber. Put deionized water (the eluent) in a 400-mL beaker to a depth of only about 0.5 cm, line the beaker with a 9-cm disc of filter paper, and cover the beaker with aluminum foil. Leave it to equilibrate while preparing the TLC sheet.

2. On a 4 x 10-cm sheet of silica gel on polyester (with fluorescent indicator), very lightly indicate (with pencil) three evenly spaced dots 1 cm above a narrow end (to keep the spots above the eluent in the developing chamber).

3. Apply about 5 µL of each of these provided solutions: (left to right) **a**, authentic adenine in the first lane; **b**, hydrolyzed plant material in the second lane; **c**, authentic cytosine in the third lane; and **d**, authentic thymine in the fourth lane. Do not apply the entirety of the 5-µL samples all at once; touch the applicators down lightly and quickly so that the spots will be kept small in diameter.

4. Place the TLC sheet in the developing chamber, reapply the cover, and allow the water eluent to climb to no higher than 1 cm from the top of the sheet. Then take the sheet out and *immediately mark the solvent front.*

5. Dry the plate under an infrared bulb or in a 100°C oven. Don't heat it longer than necessary.

6. View the chromatogram under a 254-nm ultraviolet light (**through a glass screen to protect eyes**), lightly marking the edges of the spots with a pencil. Record the distances traveled by the solvent front and by each of the spots (measure to their centers), and then calculate the R_f values (see Experiment 4).

7. To detect and find the likely identity of any sugar present, hold the TLC sheet by its upper edge and lightly paint (with a soft-bristled brush or a tuft of cotton *held in tongs*) the bottom half of the sheet with the detection reagent,[**] 0.1 M *p*-anisidine hydrochloride in 1-butanol.

---

[*] If the "exotic" plant were real, its DNA would contain four bases, two of purine type and two of pyrimidine type.
[**] Hough, L.; Jones, J. K. N.; Wadman, W. H. *J. Chem. Soc.* **1950**, 1702.

Then heat the TLC plate in a 100°C oven for 10 minutes to develop the color(s). Note and record the color, distance traveled, and R_f value for any colored spot(s). Different sugars give different colors, e.g., aldopentoses red-violet, aldohexoses green-brown, ketohexoses lemon yellow, and 2-deoxyaldoses, pale brown.

8. Now, under ultraviolet light, view any spots revealed by the detection reagent to check for fluorescence or change of spot color. A spot from a 2-deoxyaldose is expected to appear fluorescent white under UV light. Record your observation.

9. **Disposal:** Discard the TLC sheet as directed by your instructor.

II. DISCHE TEST FOR PRESENCE OF DEOXYNUCLEOSIDE UNITS

1. In a 6-inch test tube, put 0.05 g of ground seed material in 1 mL of water and add 2 mL of Dische reagent.* (**Corrosive**: it contains diphenylamine and sulfuric acid dissolved in glacial acetic).

2. Set the test tube into boiling water in a 400-mL beaker; stir and heat the contents for 10 minutes. Then observe and record any change; appearance of a blue to blue-gray color is a positive test.

3. **Disposal:** Pour the test mixture in the "Waste Organics" container.

---

* Dische, Z.; Schwarz, K. *Microchim. Acta* **1937**, 2, 13.

Nucleosides

Prelab Questions for Experiment 19

1. What makes a nucleoside a purine type?

2. Could both RNA and DNA samples give a positive Dische test? Why?

3. If sucrose is hydrolyzed and the resulting monosaccharide mixture is studied by thin layer chromatography,

    a. How many spots would be seen on the TLC plate?

    b. What colors would they be if detected using the *p*-anisidine hydrochloride detection reagent?

4. In paper or thin layer chromatography, when the spots obtained are viewed under ultraviolet light they sometimes absorb the light and appear dark. What else do they sometimes do that makes them noticeable? (Some spots don't do either of these things and simply are not detected under UV light.)

5. What is the inorganic compound that is obtained when either DNA or RNA are hydrolyzed? [It is derived from the links between nucleoside units.]

Nucleosides

Report for Experiment 19

I. CHROMATOGRAPHY

1. Draw your TLC plate here, showing the approximate positions of the spots in each lane. Use solid lines for the spots revealed by UV and dotted lines for spots revealed by the *p*-anisidine hydrochloride spray.

2. Record here the distances and R_f values:

Sample	Distance	R_f	Visual appearance
Solvent (water) front			
Adenine			
Hydrolyzed plant material			
Cytosine			
Thymine			

3. Which nitrogen bases are present in the nucleosides of this hypothetical plant?

4. Describe the color(s) and meanings of the spot(s) observed on use of the sugar detection reagent.

5. What is the structure of *p*-anisidine?

Is it a primary, secondary, or tertiary amine?

II. DISCHE TEST FOR PRESENCE OF DEOXYNUCLEOSIDE UNITS

1. Draw the structure of diphenylamine.

Is it a primary, secondary, or tertiary amine?

2. Describe the visual aspects of the test.

3. What is the meaning of the experimental result?

Related Questions for Experiment 19

1. If there were really only two nitrogen bases available for DNA, what impact would that have?

2. Thymine is one of the pyrimidines found in nucleosides. Compare uridine with thymine: where is each of them found?

3. Why are primed numbers used to denote the carbons of the sugar unit of a nucleoside?

4. a. The sugar units in the nucleotides that are linked together in DNA and RNA are said to be furanosyl forms. What does that mean?

   b. Also, the sugar units in DNA and RNA are said to be β-furanosyl forms. What does that mean?

5. Draw a pyranose form of D-ribose.

DNA and RNA*

Reference: *General, Organic, and Biological Chemistry: An Integrated Approach, 4th ed., Chapter 13.*

Purpose: With special sets of molecular models you will simulate the replication of DNA, the transcription of DNA to mRNA, and the translation of mRNA in protein synthesis.

Introduction

A detailed discussion of the replication of DNA, transcription DNA to mRNA, and the translation of mRNA is provided in your text. The following discussion is intended only to review some of the more important aspects of DNA and RNA functions as applied to your DNA kit.

THE REPLICATION OF DNA

Deoxyribonucleic acid (DNA) is the hereditary molecule in all cells. The double helix structure of DNA has a backbone consisting of a five-carbon sugar (2-deoxy-D-ribose) and phosphate groups connected by ester linkages. Attached to the backbone are four nitrogen bases. The sequence of the four bases is the genetic code, the hereditary information of the cell. The bases of one strand of the double helix are hydrogen bonded to the bases of the other strand in a complementary manner. The base thymine (T) is always hydrogen bonded to adenine (A); cytosine (C) is always hydrogen bonded to guanine (G).

A **nucleoside** is the combination of a sugar with a nitrogen base. A sugar, a nitrogen base, and a phosphate group combine to form a **nucleotide**. DNA and RNA are polymers of nucleotides in which the sugars are 2-deoxy-D-ribose and D-ribose, respectively.

When a cell divides, each daughter cell receives DNA identical to the parent cell through the process of **replication**. During replication of DNA, the double helical structure of DNA unwinds and each strand serves as the pattern for the synthesis of a new complementary strand of DNA.

$$\text{DNA} \xrightarrow{\text{Replication}} \text{DNA} + \text{DNA}$$

(in parent cell)   (in two daughter cells)

The DNA model kit used in this experiment consists of interlocking brightly colored plastic components, representing the following structures: phosphates, five-carbon sugars, nitrogen bases,

---

*The DNA kits used in this experiment were designed by Ken Zietlow and Richard Brady and are distributed by Ward's Natural Science, 5100 West Henrietta Road; P.O. Box 92912; Rochester, NY 14692-9012; Product No. 81 V 0140; www.wardsci.com. This experiment is based on a series of exercises in an accompanying *Student Guide for DNA Model Kit* and is published by permission of the authors.

amino acids, hydrogen, and hydroxyl groups. An itemized list of pieces is provided in Part I, Step 2 of the Procedure section. Ester linkages are represented by tightly interlocked pieces. Hydrogen bonding between bases is denoted by loose-fitting, but not interlocking, complementary structures.

In Part I of this experiment you will construct a model based upon a specific base sequence in a DNA strand. You will then assemble a second complementary strand of the DNA structure, linking together the bases that hydrogen bond to one another. The two strands will then be separated, and each of the two daughter strands will be used to assemble new DNA structures.

TRANSCRIPTION OF DNA TO RNA

Each cell must synthesize protein and each protein must have a specific order of amino acids (the primary structure of the protein). The sequence of nitrogen bases in DNA contains the genetic information required to assemble the amino acids in the correct order. Another nucleic acid, **ribonucleic acid (RNA)**, copies the genetic information from DNA and uses that information to direct the synthesis of protein in the cell's ribosomes. Like DNA, RNA has a backbone made of a five-carbon sugar with ester linkages to phosphate groups; however, the sugar in RNA is ribose rather than deoxyribose. Four nitrogen bases are attached to the backbone; three of the bases (adenine, cytosine, and guanine) are the same as in DNA. The RNA base uracil replaces the DNA base thymine. Several types of RNA exist, usually with a single-strand of nucleic acid.

The preliminary step in protein synthesis is a **transcription** of the genetic information from a segment of DNA to **messenger RNA (mRNA)**. This step is similar to the replication of DNA but only a segment of the DNA chain is copied, not the entire chain.

In Part II of this experiment, you will use models to demonstrate transcription of DNA to mRNA. During transcription the nitrogen bases of DNA hydrogen bond to those of mRNA. DNA adenine (A) hydrogen bonds to *uracil* (U) in the growing RNA, DNA guanine to cytosine (C) in the RNA, DNA cytosine (C) to guanine (G) in the RNA, and DNA thymine (T) to adenine (A) in the RNA.

THE TRANSLATION OF mRNA (PROTEIN SYNTHESIS)

In the preliminary step of protein synthesis, genetic information was transcripted from DNA to mRNA in the nucleus of the cell. After being synthesized, the mRNA migrates to the cell's ribosomes.

Another form of RNA, **transfer RNA (tRNA)**, is much smaller than other nucleic acids. Each tRNA molecule is specifically designed for a particular amino acid. The amino acid becomes attached to the end of its tRNA molecule. One after another, tRNA molecules hydrogen bond to mRNA; the amino acids carried by the tRNA are assembled in sequence to make a protein. The order of the amino acid assembly is determined by the sequence of nitrogen bases in the mRNA. A sequence of three nitrogen bases on mRNA is a **codon**. The nitrogen bases of the mRNA codon hydrogen bond to complementary nitrogen bases in the **anticodon** of the tRNA. Most of the 20 common amino acids have more than one codon. For example, UUU and UUC are both codons for the amino acid phenylalanine.

To summarize, the preliminary step of protein synthesis, **transcription**, copies genetic information from DNA to RNA. The actual synthesis of protein occurs during **translation** when the sequence of nitrogen bases is translated into the sequence of amino acids of the protein. The key to the translation is the matching of specific amino acids to the codons of the mRNA using complementary anticodons on tRNA.

Part III of this experiment illustrates the repeated steps that take place between tRNA and mRNA in synthesizing a polypeptide (protein).

Procedure

I. THE REPLICATION OF DNA

1. Obtain a DNA kit from your instructor. Work in pairs because the components of two kits will be needed for this experiment.

2. Separate the kit into its components by letter. You should have the following parts in each kit:

Symbol	Quantity	Part Name
A	24	Deoxyribose
B	12	Ribose
C	24	Phosphate
D	6	Uracil (U)
E	6	Thymine (T)
F	6	Guanine (G)
G	6	Cytosine (C)
H	2	Hydrogen
J	6	Adenine (A)
OH	2	Hydroxyl group
Amino acids	2	Two different amino acids of different color

3. Remove one each of Parts D, E, F, G, and J (uracil, thymine, guanine, cytosine, and adenine).

4. Test to see which **pairs** of the above parts hydrogen bond by attempting to join the pieces together as in a jigsaw puzzle. Try every combination possible.

5. With the use of parts from two kits, use A, C, E, F, G, and J to assemble 48 separate DNA nucleotides. Remember that each nucleotide consists of a deoxyribose sugar, a phosphate, and a nitrogen base. See Figure 20-1.

Quantity	DNA Nucleotides
12	phosphate + thymine + deoxyribose
12	phosphate + adenine + deoxyribose
12	phosphate + guanine + deoxyribose
12	phosphate + cytosine + deoxyribose

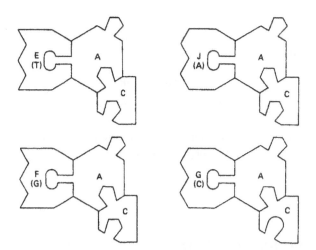

Figure 20-1 DNA molecules.

6. Arrange all nucleotides so that the phosphate groups (C) are on the right side of deoxyribose (A).

7. Attach 12 of these nucleotides end to end so that the sequence of bases matches the diagram of Figure 20-2. The letter symbol is that of the bases, not the code symbol for each part. See Step 2.

Figure 20-2 Base sequence in a DNA strand.

8. The 12 remaining nucleotides will now be used to construct the second strand of the DNA molecule. Begin linking together the bases that hydrogen bond and then attach the deoxyribose and phosphate structures of all nucleotides. You should now have a ladderlike arrangement, which represents the double-stranded DNA.

9. Separate the two strands of the DNA molecule, keeping each strand intact.

10. Using the 24 remaining nucleotides, construct two DNA molecules as you did earlier in Step 8. You have simulated the replication of DNA.

II. TRANSCRIPTION OF DNA TO mRNA

1. Disassemble one of the DNA molecules that you constructed in Part I. Completely separate the individual pieces (sugars, phosphates, and bases), and group similar molecules together.

2. Assemble 24 RNA nucleotides as indicated below. Make sure that the phosphate groups are on the right side of the ribose. See Figure 20-3. These nucleotides will constitute the RNA molecule pool in the cell.

Quantity	RNA Nucleotides
6	phosphate + uracil + ribose
6	phosphate + guanine + ribose
6	phosphate + adenine + ribose
6	phosphate + cytosine + ribose

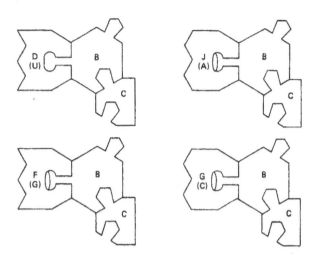

Figure 20-3 RNA nucleotides.

3. Construct a tRNA molecule by linking three nucleotides together as illustrated in Figure 20-4. The nucleotide should contain the bases GUA.

4. Construct three other tRNA molecules containing the bases GAA, CCU, and UGC.

5. Attach the amino acids, with H and OH on each end, to each of the four sets of tRNA molecules corresponding to the anticodons listed below. See Figure 20-4. Place the molecules aside for later use.

Anticodon	Amino acid
GUA	histidine
GAA	leucine
CCU	glycine
UGC	threonine

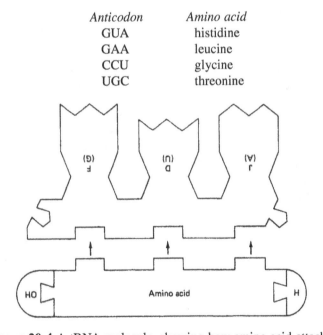

Figure 20-4 A tRNA molecule, showing how amino acid attaches.

6. Separate the double-stranded DNA as before. With the remaining nucleotides that you have constructed in Step 2, match the bases of the RNA nucleotides with the original strand of DNA. Remember, "matching bases" form hydrogen bonds.

7. Connect the RNA nucleotides. You have now constructed an mRNA molecule.

8. Place the mRNA molecule to one side and attach the two strands of DNA as before. The DNA molecule is back in its original state, capable of taking part in another transcription process if needed.

III. THE TRANSLATION OF mRNA (PROTEIN SYNTHESIS)

1. Position the mRNA molecule you have constructed in Part II on your laboratory table so that you can easily "match bases" as illustrated in Figure 20-5.

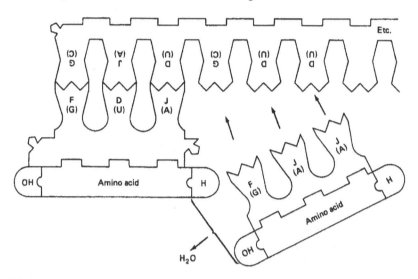

Figure 20-5 Translation of mRNA to polypeptide. The codons are written "upside down" on the mRNA molecule because the codon/anticodon pairings occur in an antiparallel direction.

2. Match anticodons on amino acid-carrying tRNAs (from Part II, Step 5) with the codons on the mRNA. You should have four amino acids placed side by side.

3. Construct a polypeptide by "splitting off" water molecules (H and OH) and attaching the amino acid fragments. Remove the polypeptide molecule. As each amino acid is removed from a tRNA and attached to the growing polypeptide, the tRNA is free to return to the amino acid pool to pick up another amino acid. This process replicates the synthesis of a polypeptide using the DNA genetic code. The sequence of amino acids in the polypeptide was determined by the sequence of nucleotides in DNA.

Prelab Questions for Experiment 20

1. List the nitrogen bases found in DNA.

   _____

   _____

   _____

   _____

2. List the nitrogen bases found in RNA.

   _____

   _____

   _____

   _____

3. How is an ester linkage identified in the models used in this experiment?

   _____

   _____

4. How is a hydrogen bond identified in the models used in this experiment?

   _____

   _____

5. Which part of the experiment illustrates the synthesis of a protein?

   _____

Report for Experiment 20

I. THE REPLICATION OF DNA

    1. List the nitrogen base pairs that hydrogen bond.

_____

_____

_____

_____

    2. List the base sequence in the second DNA strand.

    3. a.   Does your DNA model represent the true geometry of the DNA molecule? _____

        b.   If not, what is the major difference in structure?

_____

_____

II. TRANSCRIPTION OF DNA TO mRNA

    1. List the base sequences on the mRNA molecule of Step 7.

    2. What are the three-base sequences on the mRNA molecule called?

_____

III. THE TRANSLATION OF mRNA (PROTEIN SYNTHESIS)

1. Using the standard method of illustrating polypeptide structure (e.g., Ala-Glu-Ser-Val), write the polypeptide formed in this experiment.

   _____

2. What are the three-base sequences on the tRNA called?

   _____

3. What happens to the mRNA after the polypeptide is synthesized?

   _____

   _____

Related Questions for Experiment 20

1. Suppose that the base sequence for one strand of DNA in Part I of this experiment was as follows:

   a. List the base sequence in the second DNA strand.

$$\overline{\underset{|}{G}\ \underset{|}{A}\ \underset{|}{A}\ \underset{|}{G}\ \underset{|}{T}\ \underset{|}{A}\ \underset{|}{T}\ \underset{|}{G}\ \underset{|}{C}\ \underset{|}{C}\ \underset{|}{C}\ \underset{|}{T}}$$

   b. List the base sequence in the mRNA strand that is formed from the DNA strand shown above.

   c. List the codons in the mRNA strand of Question 1b.

   _____   _____   _____   _____

   d. List the anticodons in the four tRNA molecules that would hydrogen bond to the mRNA of Question 1b.

   _____   _____   _____   _____

   e. Using the standard method of illustrating polypeptide structure, write the polypeptide that would be formed on the mRNA strand from Question 1b.

   _____

2. a. What is a mutation?

   _____

   _____

b. Give at least two examples of how a mutation can occur.

_____

_____

_____

_____

3. a. Must two mRNA molecules be identical to produce the same polypeptide?

_____

b. Explain.

_____

_____

_____

4. With the help of reference sources, list some of the diseases associated with mutations, and briefly describe their symptoms.

_____

_____

_____

_____

_____

_____

_____

_____

_____

_____

DNA and RNA

Appendix A
Exercise on Chemical Periodicity

Reference: *General, Organic and Biological Chemistry: An Integrated Approach, 4th ed., Chapters 2, 3, and 4. See also Table of Elements and Periodic Table inside front and back covers.*

Purpose: To provide additional opportunities to study the structure of the periodic table and use the Periodic Law to predict properties of elements, properties of compounds, and formulas of compounds.

Introduction

HISTORICAL BACKGROUND

During the eighteenth and nineteenth centuries chemists began to recognize similarities in the properties of some of the elements. They found, for example, that calcium, barium, and strontium could be isolated by similar chemical methods. Sulfur, selenium, and tellurium formed similar chemical compounds, and iodine, bromine, and chlorine were colored nonmetallic elements.

By the early 1800s, the atomic masses (atomic weights) of some of the elements were determined with a fair degree of accuracy. In 1817, Johann Wolfgang Döbereiner made an interesting observation. He noted that in several instances, when three elements having similar chemical and physical properties were arranged in ascending order of atomic masses, the mass of the middle element was almost equal to the average of the first and third. For example consider the similar elements chlorine (atomic mass 35.5), bromine (atomic mass 79.9), and iodine (atomic mass 127). The average of the atomic masses of chlorine and iodine is $(127 + 35.5) \div 2 = 81.3$, fairly close to the accepted atomic mass of bromine (79.9).

Similarly, the atomic mass of strontium, 87.6, is close to the average of the atomic masses of calcium and barium: $(40 + 137) \div 2 = 88.5$. A third such relationship exists in the case of selenium (atomic mass 79.2)—not too different from the average atomic masses of sulfur and tellurium: $(32 + 127.6) \div 2 = 79.7$.

Döbereiner's system of **Triads** was the first systematic attempt to classify the chemical elements but Döbereiner's contemporaries did not pay much attention to his work. Today we realize that Döbereiner's Triads are three elements in the same chemical family. See Figure A-1.

In 1864, the English chemist John Newlands grouped all the known elements of his day into groupings of seven elements each. He observed that when the elements were arranged in ascending order of atomic mass, the chemical properties of the elements reappeared every eighth element, just as the notes on a musical scale reappear every eighth note. For example, the eighth element after lithium was sodium; lithium and sodium have similar properties. (The noble gases were not discovered until later that century.)

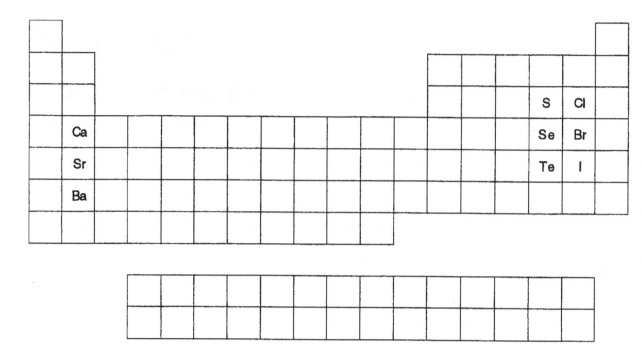

Figure A-1 Some of Döbereiner's Triads

In a modern version of Newlands' brief table, shown in Figure A-2, elements are arranged in order of increasing atomic masses. Elements aligned vertically have similar physical and chemical properties. Newlands' **Law of Octaves** didn't work well beyond the element calcium and his work was not well accepted by other chemists who ridiculed his comparison of elements to a piano keyboard. One critic even sarcastically said arranging the elements in alphabetical order would be as useful as in order of atomic mass. His critics were wrong.

Li	Be	B	C	N	O	F
6.9	9.0	10.8	12.0	14.0	16.0	19.0
do	re	me	fa	so	la	ti
Na	Mg	Al	Si	P	S	Cl
23.0	24.3	27.0	28.1	31.0	32.1	35.5
do	re	me	fa	so	la	ti

Figure A-2 Newlands' Law of Octaves arranged elements in order of increasing atomic masses. Properties repeat themselves like notes on a musical scale.

MENDELEEV'S PERIODIC TABLE

The Periodic Table always existed in nature; Newlands and others were groping toward the Periodic Law but were not yet aware of it. Lothar Meyer, in Germany, and Dmitri Mendeleev, in Russia, independently discovered the Periodic Law and created the periodic table of the elements. Meyer did not publish his work; Mendeleev did in 1869 and received more credit. Unlike Döbereiner and others who sought mathematical relationships among the elements, Mendeleev and Meyer looked for, and found, trends in properties of the elements.

All 63 known elements in Mendeleev's periodic table were arranged in horizontal rows of increasing atomic masses. Because his classification revealed recurring patterns in the elements, Mendeleev was able to leave spaces in his table for elements that he correctly predicted would be discovered. His table gained increased acceptance when three predicted elements were subsequently discovered. The first of these was gallium, discovered in 1875. See Figure A-3. Mendeleev even correctly predicted how the element would be discovered—with a spectroscope.

	Mendeleev's Predicted Properties for Gallium	Actual Properties of Gallium
Atomic mass	68	69.7
Density	5.9	5.903
Melting point	low	29.8°C
Formula of oxide	Ga_2O_3	Ga_2O_3

Figure A-3 Comparison of predicted and experimentally determined properties of gallium.

Mendeleev also argued for liberal, social, and political reforms in Russia and was forced to leave the University of Saint Petersburg. Element 101, mendelevium, was named to honor his contribution to chemistry.

Mendeleev's table contained a few elements that appeared to be out of order when arranged by atomic masses so Mendeleev assumed the atomic masses were incorrect and arranged several elements by their properties rather than their atomic masses. Other elements discovered later—the lanthanide series—had no places on Mendeleev's table.

Englishman Henry Gwyn-Jeffreys Moseley discovered the concept of atomic number in 1912. This discovery allowed him to make a new periodic table, one arranged in order of increasing atomic numbers. His new table cleared up the remaining shortcomings of Mendeleev's table. Moseley's table left gaps for seven missing elements—these were later discovered. His experimental work was a remarkable achievement. In 1915, Mosely died at age 28 while fighting in World War I.

MODERN PERIODIC LAW

The modern **Periodic Law** states that the properties of the elements repeat themselves periodically if the elements are arranged in order of increasing atomic numbers. Many different versions of a periodic table have been developed on the basis of this law. The most common periodic table, shown on the inside front cover, arranges similar elements in vertical columns called **groups**. The horizontal rows are called **periods**. The location of an element in the table can be used to predict its physical and chemical properties. The discovery of the Periodic Law is one of the most useful achievements in science. The Law is an extremely powerful tool to chemists, one that all chemists need to master.

Each chemical group is identified by a number of reference. Unfortunately, two rival numbering systems have existed for many years. In the United States, elements in the scandium group were labeled in Group 3B (previously IIIB) and elements in the boron group were labeled 3A (previously IIIA). Elsewhere, most chemists followed the "official" International Union of Pure and Applied Chemistry (IUPAC) system which reversed the IIIA and IIIB designations. In an attempt to resolve this confusion, new group numbers were proposed that put scandium in Group 3 and boron in Group 13. These two numbering systems are shown in Figure A-4. We will use the system that designates the columns with a number and letter. Most chemists are not concerned with what the groups are called so long as everyone is consistent. U.S. chemical educators, however, prefer that boron be in Group 3A. It is easier to explain to students that the representative elements are the "A" groups, and that for any **representative element**, the group number equals the number of valence electrons. Thus, boron, aluminum, and the rest of Group 3A all have three valence electrons.

Ninety of the elements occur naturally on earth, some in extremely minute quantities. Technetium (atomic number 43), promethium (61), and the transuranium elements (atomic numbers greater than 92) are man-made. At the time of this writing, elements through atomic number 112 plus 114 and 116 have been synthesized. Preparation of the remaining elements of Period 7 has been claimed but not substantiated. Several more new elements are likely to be made in the next year or two because of recent technological advances. With each new element scientists learn more about matter and the atom. Ultimately, this work is concerned with how the world was made. Initially no practical uses of these new elements are foreseen. But significant uses are possible someday, just as important uses were eventually found for man-made elements such as technetium, plutonium, americium, and californium.

Some elements in Period 7 have temporary names and symbols, to be used until the existence and the discovery of the element can be verified and the official name determined by IUPAC. Official names have recently been assigned to 112 (copernicum, Cn), 114 (flerovium, Fl), and 116 (livermorium, Lv).

One shortcoming of the usual form of the periodic table is that it does not show the true position of the **lanthanide series** or rare earth elements (elements 58-71) and the **actinide series** (elements 90-103) of elements. The long form of the periodic table, Figure A-4, shows these elements in their correct positions.

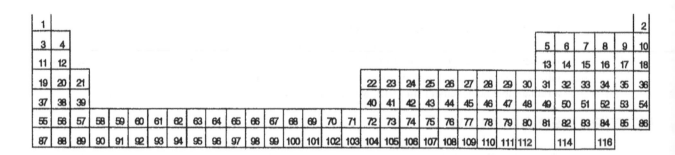

Figure A-4 The long form of the periodic table of the elements.

In any form, the periodic table and the Periodic Law are useful in predicting such things as

- The properties of elements, such as size of atoms and electronegativity.
- The properties of compounds formed from the elements.
- The formulas of compounds.

PROPERTIES OF THE ELEMENTS

Classification

Note on the Periodic Table shown on the inside back cover the very useful categorization of the elements into **metals**, **semimetals**, and **nonmetals**. Metals usually react by giving up outer shell electrons, nonmetals by gaining electrons, and semimetals have an intermediate nature.

Atomic Radius

An atom can be thought of as a tiny sphere with the nucleus in the center. When two atoms are near each other, the electrons from atom 1 are attracted to the protons of atom 2, and the protons from atom 1 are repelled by the protons from atom 2. The two atoms find a balance between this attraction and repulsion and end up a certain distance apart. When two like atoms are bonded to each other, the **atomic radius** is half the distance between the two nuclei. Figure A-5 shows the atomic radii of the elements; the trends in atomic radius are more apparent when displayed graphically in Figure A-6. The units of the radii are Ångstroms, a non-SI metric unit. One Å equals 10^{-10} meter. When the elements are arranged in order of atomic number, the atomic radius shows a trend that repeats itself. This trend is evident only when the elements are arranged in order of atomic number; arranging the elements in alphabetical order would not reveal the trend. The trend can be used to predict the radius of undiscovered elements.

H 0.32																	He 0.93
Li 1.23	Be 0.90											B 0.82	C 0.77	N 0.75	O 0.73	F 0.72	Ne 0.71
Na 1.54	Mg 1.36											Al 1.18	Si 1.11	P 1.06	S 1.02	Cl 0.99	Ar 0.98
K 2.03	Ca 1.74	Sc 1.44	Ti 1.32	V 1.22	Cr 1.18	Mn 1.17	Fe 1.17	Co 1.16	Ni 1.15	Cu 1.17	Zn 1.25	Ga 1.26	Ge 1.22	As 1.20	Se 1.16	Br 1.14	Kr 1.89
Rb 2.16	Sr 1.91	Y 1.62	Zr 1.45	Nb 1.34	Mo 1.30	Tc 1.27	Ru 1.25	Rh 1.25	Pd 1.28	Ag 1.34	Cd 1.41	Sn 1.44	Sn 1.41	Sb 1.40	Te 1.36	I 1.33	Xe 1.31
Cs 2.35	Ba 1.98	La 1.25	Hf 1.44	Ta 1.34	W 1.30	Re 1.28	Os 1.26	Ir 1.27	Pt 1.30	Au 1.34	Hg 1.49	Tl 1.48	Pb 1.47	Bi 1.46	Po 1.53	At 1.47	Rn -
Fr -	Ra -	Ac -															

Ce 1.65	Pr 1.65	Nd 1.64	Pm 1.63	Sm 1.62	Eu 1.85	Gd 1.61	Tb 1.59	Dy 1.59	Ho 1.58	Er 1.57	Tm 1.56	Yb 1.70	Lu 1.56
Th 1.65	Pa -	U 1.42	Np -	Pu 1.08	Am -	Cm -	Bk -	Cf -	Es -	Fm -	Md -	No -	Lr -

Figure A-5 Atomic radii of the elements in units of Ångstroms.

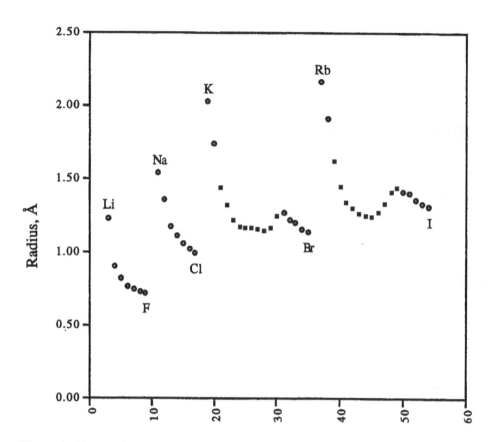

Figure A-6 Trends in atomic radius are more obvious when graphed. In order to make the trends more apparent, hydrogen and the noble gases are not included. Transition elements are represented by ■.

Electronegativity

All atoms (except the noble gases) have fewer than eight valence electrons. **Valence electrons** are the electrons in the valence shell (the highest occupied energy level). For the representative elements, the number of valence electrons is equal to the group number; for example carbon is in Group 4A and therefore has four valence electrons. The nucleus of all atoms (except the noble gases) has a tendency to attract electrons from outside the atom. **Electronegativity** is the power of an atom to attract electrons towards itself when it is part of a compound. Electronegativity values are numbers from a scale originally set up by Nobel Laureate Linus Pauling and later slightly modified by others. Figure A-7 shows trends in electronegativities among the elements; these trends are more apparent when graphically displayed in Figure A-8.

It will be very helpful in predicting ionic vs. covalent bonding and bond polarity to note that there are approximately 0.5 unit increases in electronegativity on moving across the elements of the Period 2 from lithium at 1.0 to fluorine at 4.0 and that electronegativity decreases as you go down the elements of each group. Additionally, because hydrogen combines with virtually all other elements, it is important to know its electronegativity, 2.1.

H																	He
2.1																	
Li	Be											B	C	N	O	F	Ne
1.0	1.6											2.0	2.5	3.0	3.4	4.0	
Na	Mg											Al	Si	P	S	Cl	Ar
0.9	1.3											1.6	1.9	2.2	2.6	3.2	
K	Ca	Sc	Ti	V	Cr	Mn	Fe	Co	Ni	Cu	Zn	Ga	Ge	As	Se	Br	Kr
0.8	1.0	1.4	1.5	1.6	1.7	1.6	1.8	1.9	1.9	1.9	1.7	1.8	2.0	2.2	2.6	3.0	
Rb	Sr	Y	Zr	Nb	Mo	Tc	Ru	Rh	Pd	Ag	Cd	Sn	Sn	Sb	Te	I	Xe
0.8	1.0	1.2	1.3	1.6	2.2	1.9	2.2	2.3	2.2	1.9	1.7	1.8	2.0	2.1	2.1	2.7	
Cs	Ba	La	Hf	Ta	W	Re	Os	Ir	Pt	Au	Hg	Tl	Pb	Bi	Po	At	Rn
0.8	0.9	1.1	1.3	1.5	2.4	1.9	2.2	2.2	2.3	2.6	2.0	2.0	2.3	2.0	2.0	2.2	
Fr	Ra	Ac															
0.7	0.9	1.1															

Ce	Pr	Nd	Pm	Sm	Eu	Gd	Tb	Dy	Ho	Er	Tm	Yb	Lu
1.1	1.1	1.1	1.1	1.2	1.2	1.2	1.1	1.2	1.2	1.2	1.3	1.1	1.3
Th	Pa	U	Np	Pu	Am	Cm	Bk	Cf	Es	Fm	Md	No	Lr
1.3	1.5	1.4	1.4	1.3	1.3	1.3	1.3	1.3	1.3	1.3	1.3	1.3	1.3

Figure A-7 Electronegativities of the elements.

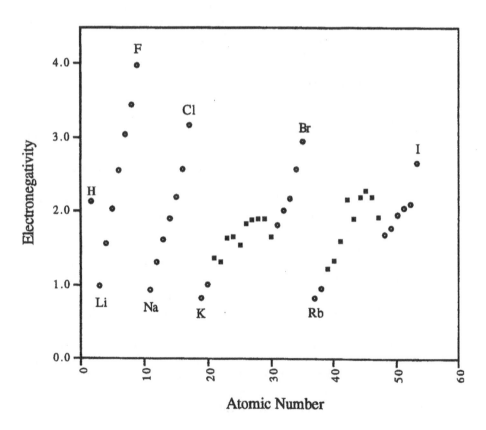

Figure A-8 Trends in electronegativities of the elements. In order to make the trends more apparent, hydrogen and the noble gases are not included. Transition elements are represented by ■.

PROPERTIES OF COMPOUNDS

Chemical compounds generally fall into one of two broad categories: ionic compounds and covalent compounds.

Ionic compounds are always solid at room temperature. They have high melting points, over 350°C and often as high as 1000°C. Most ionic compounds dissolve in water; few dissolve in nonpolar solvents like paint thinner. Ionic compounds always conduct an electric current when dissolved in water or when heated to the liquid state. They do not burn and few have odors.

Covalent compounds are also called molecular compounds. Although some are solid at room temperature, many are liquids or gases. Their melting points are low—less than 350°C. A few covalent compounds dissolve in water; many dissolve in nonpolar solvents. Few conduct an electric current. Most will burn and many have pronounced odors.

It is possible to predict whether a compound will have properties of an ionic or a covalent compound. The most accurate prediction requires calculating the difference in electronegativity between two atoms bonded to each other. When the difference in electronegativity values is large (1.9 or greater), the major attraction between the atoms is an ionic bond. When the difference in electronegativity values is small (less than 1.9), the major attraction between the atoms is a covalent bond. Most bonds have a mixture of ionic and covalent character but one character predominates and it determines the properties of the compound.

A faster but slightly less accurate prediction can be made by first classifying the elements as metals or nonmetals. Metals, as seen in Figure A-7, have low electronegativities; nonmetals have high electronegativities. A bond between a metal and a nonmetal will therefore generally have a large difference in electronegativities and will be an ionic bond. A bond between a nonmetal and a nonmetal will generally have a low difference in electronegativities and will be a covalent bond.

FORMULAS OF COMPOUNDS

The **Octet Rule** (G.N. Lewis, 1916) states that atoms tend to gain, lose, or share electrons in order to end up with the same electron configuration as one of the noble gases. All the noble gases have eight valence electrons—except for helium, which has two valence electrons. Thus, atoms tend to gain, lose, or share electrons in order to end up with eight valence electrons. Hydrogen is the exception to the rule of eight; this nonmetal is usually placed above the alkali metals on the periodic table because it, too, has one valence electron. Like the other Group I elements, hydrogen can exhibit the behavior typical of metals and lose an electron, forming a positively charged ion (a **cation**). Hydrogen ions are remarkable in that, accurately speaking, they are a bare atomic nucleus, consisting only of the subatomic particle called a proton. This is such a high-energy specie that it can only exist when associated with electron-rich molecules, most commonly water molecules, and then is represented as the hydronium ion, H_3O^+. But, consistent with its much higher electronegativity than the other elements in Group I, hydrogen can alternatively complete its outer shell of electrons by adding or sharing to reach its full complement of two.

Metal atoms can satisfy the Octet Rule by losing one or more electrons to become positively charged ions or cations. It is easy to predict the charge of the ions of the alkali metals, the alkaline earth metals, aluminum, and a few other metals that always form the same ion. The group number of the representative elements is the number of valence electrons; metals will lose all their valence electrons. The group number of representative metals is the positive charge of the ion.

Group	Charge of Ion	Example	
1A	1+	Na^+	sodium ion
2A	2+	Ca^{2+}	calcium ion
3A	3+	Al^{3+}	aluminum ion

Nonmetal atoms can satisfy the Octet Rule in two ways: (1) by sharing electrons with another nonmetal to form a covalent bond, or (2) by gaining one or more electrons to form negatively charged ions or **anions**. It is easy to predict the charge of nonmetal anions. The group number is the number of valence electrons; subtract the group number from 8 to find the negative charge of the nonmetal monoatomic ion. Names of nonmetal monoatomic ions always end with an **-ide** suffix.

Group	Charge of Ion	Example	
5A	3–	N^{3-}	nitride ion
6A	2–	O^{2-}	oxide ion
7A	1–	Cl^-	chloride ion

Binary compounds contain only two elements and their formulas are easily predicted using the periodic table. Consider the compound calcium chloride which is sometimes used to melt ice on sidewalks. Calcium is a metal in Group 2A, therefore calcium has two valence electrons. By applying the Octet Rule, we see calcium must lose its two valence electrons and so calcium has a 2+ charge, Ca^{2+}. The nonmetal chlorine is in Group 7A and has seven valence electrons. Chlorine atom needs to gain one electron to satisfy the Octet Rule and so chloride ion has a charge of 1–, Cl^-.

Once you've used the periodic table to predict the charge of the metal cation and the nonmetal anion, list the positive ion first and the negative ion second. Then simply "criss-cross" the charges to determine the subscripts of the compound's formula. The charge of calcium (2) becomes the subscript of chlorine. The charge of chlorine (1) becomes the subscript of calcium. Subscripts of "1" are not written; when an atom does not have a subscript it means there is one atom of that element. So instead of Ca_1Cl_2, calcium chloride's formula is $CaCl_2$.

This "criss-cross" method works in all cases except where the subscripts are equal. For example, criss-crossing the charges of calcium ion (Ca^{2+}) and oxide ion (O^{2-}) gives the incorrect formula Ca_2O_2. Formulas of ionic compounds are always written as simplest whole-number ratios. For calcium oxide, the simplest ratio is not 2 to 2, but 1 to 1. The correct formula is CaO.

The names of binary ionic compounds list the metal name first, followed by the nonmetal name with an -ide suffix.

As stated above, when the difference in electronegativities of two bonded atoms is small, they form covalent bonds. Lewis explained this as a sharing of electrons. You can predict the number of covalent bonds (single, double, or triple) from a compound's molecular formula and the number of valence (outer shell) electrons available in the atoms of the component elements by application of the Octet Rule. See Chapter 3 in the Raymond, 4th ed., text.

Name _____ Section _____ Date _____

PRELIMINARY QUESTIONS

1. Why is the format of the periodic table shown inside the back cover used more frequently than the long form of the periodic table in Figure A-4?

   _____

   _____

2. What is the principal difference between the arrangement of elements on Mendeleev's periodic table and our present periodic table?

   _____

   _____

3. Complete the following table by comparing the properties of ionic and covalent compounds.

	Ionic Compounds	*Covalent Compounds*
Melting Point		
Solubility in Water		
Conduction of Electric Current		
Flammability		
Odor		

4. Naphthalene has a melting point of 80.5°C, is soluble in nonpolar solvents but not water, and does not conduct an electric current. Is naphthalene likely an ionic compound or a covalent compound?

   _____

5. List the elements in Period 2 of the periodic table.

   _____

   _____

6. List the elements in the alkali metal family (Group 1A).

   _____

   _____

7. Find the electronegativity (EN) of each atom in the following compounds. Find the difference in electronegativity (ΔEN) between the atoms in each pair. Predict whether the bond between the two atoms is ionic or covalent.

	EN1 (atom 1)	EN2 (atom 2)	$\Delta EN = EN2 - EN1$	Bond Type
K-F				
Li-Br				
I-Cl				
Cl-Cl				

8. a. What numerical difference in electronegativity results in a covalent bond?

_____

b. What numerical difference in electronegativity results in an ionic bond?

_____

9. a. In general, what type of bond is formed when a metal bonds with a nonmetal?

_____

b. In general, what type of bond is formed when a nonmetal bonds with a nonmetal?

_____

Procedures

I. STRUCTURE OF THE PERIODIC TABLE

Place your answers to these questions and those of Parts II and III in the following Report Section. Use the blank periodic table in Figure A-9 to complete the following steps. Use colored pencils to help identify the selected feature. You may use your textbook as a reference.

1. Number the periods and the groups.

2. Label the following chemical groups:
   a. Alkali metals
   b. Alkaline earth metals (Group 2A)
   c. Halogens (Group 7A)
   d. Noble gases (Group 8A)

3. Sketch a dark black line separating the metals from the nonmetals.

4. Label the following elements:
   a. Representative elements
   b. Transition elements
   c. Lanthanide series (formerly called rare earth elements)
   d. Actinide series

II. PROPERTIES OF ELEMENTS

A. Atomic Radius

1. Period 1 is not included in Figure A-6. Find the elements in Period 2 in Figure A-6 and note the trend in atomic radius as atomic numbers increase in Period 2. Find the elements in Period 3 and note the trend in Period 3. What generalization can you draw about the trend in radius as atomic numbers increase in a period?

2. Find the alkali metals in Figure A-6 and note the trend in atomic radius as atomic numbers increase in this group. Find the halogens and note the trend within that group. What generalization can you draw about the trend in atomic radius as atomic numbers increase in a group?

B. Electronegativity

1. Period 1 includes hydrogen, whose electronegativity is 2.1. In Figure A-8 find the elements in Period 2 and note the trend in electronegativity. Note the trend in Period 3. What generalization can you make about the trend in electronegativity as atomic numbers increase in a period?

2. Find the alkali metals in Figure A-8 and note the trend in electronegativity as atomic numbers increase in this group. Find the halogens and note the trend within that group. What generalization can you draw about the trend in electronegativity as atomic numbers increase in a group?

C. Charges of Monoatomic Ions

1. Using the Octet Rule, write the symbol (including charge) of the following representative element metal ions in the appropriate position in the blank periodic table of Figure A-10.
   a. Lithium ion
   b. Sodium ion
   c. Potassium ion
   d. Magnesium ion
   e. Calcium ion
   f. Strontium ion
   g. Barium ion
   h. Aluminum ion

2. Write the symbol (including charge) of the following monoatomic nonmetal ions in the appropriate position in Figure A-10.
   a. Fluoride ion
   b. Chloride ion
   c. Bromide ion
   d. Iodide ion
   e. Oxide ion
   f. Sulfide ion
   g. Nitride ion

III. BONDING TYPE, FORMULAS, AND NAMES OF COMPOUNDS

Answer the questions asked in Report Section III, using the information provided on polyatomic ions.

Report Section for Chemical Periodicity Exercise

I. STRUCTURE OF THE PERIODIC TABLE

Follow Steps 1-4 in Procedure II to fill in Figure A-9.

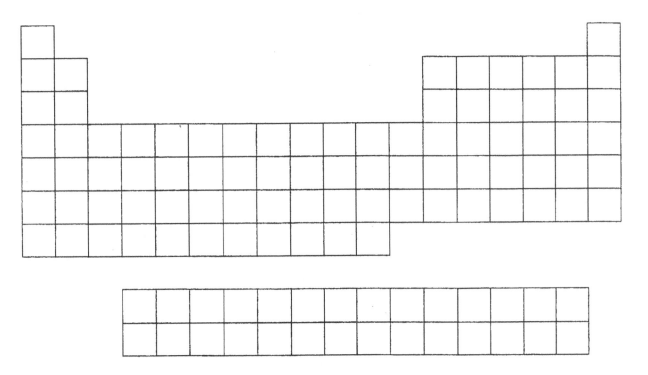

Figure A-9

II. PROPERTIES OF ELEMENTS

A. Atomic Radius

1. What generalization can you make about the trend in atomic radius as atomic number increases in a period?

   _____

   _____

2. What generalization can you make about the trend in atomic radius as atomic number increases in a group (family)?

   _____

   _____

B. Electronegativity

1. What generalization can you make about the trend in electronegativity as atomic number increases in a period?

   _____

   _____

2. What generalization can you make about the trend in electronegativity as atomic number increases in a group?

   _____

   _____

C. Charges of Monoatomic Ions

1. Follow Steps 1 and 2 in Procedure III C to fill in Figure A-10.

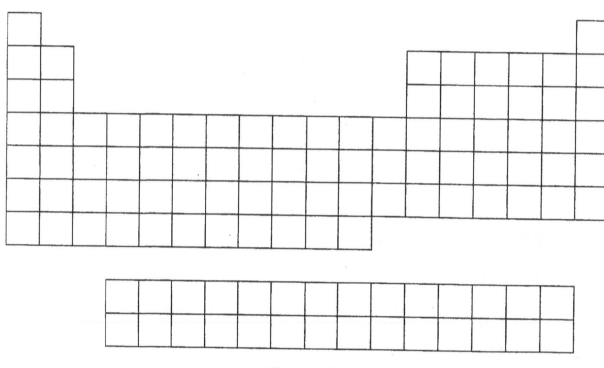

Figure A-10

2. Explain how a representative element's position on the periodic table is helpful in determining the charge of its monoatomic ion.

   _____

   _____

III. BONDING TYPE, FORMULAS, AND NAMES OF COMPOUNDS

Polyatomic anions (note that bonding within the ions is covalent) whose names and structures are important for work on the next sections:

Bicarbonate	HCO_3^-	
Carbonate	CO_3^{-2}	
Hydroxide	HO^- [or OH^-]	^-O-H
Nitrate	NO_3^-	
Phosphate	PO_4^{-3}	
Sulfate	SO_4^{-2}	

1. Classify each compound as ionic or covalent by using the electronegativity difference or the metal/nonmetal technique (if there is a metal, the compound is ionic; covalent compounds do not contain metals). If an ionic compound, write the formula and charge of each ion (some ions are monoatomic and some are polyatomic), then use the charges of the ions to predict the formula of the ionic compound. If a covalent compound, do not write the formulas and charges of ions; instead, place a large X in the boxes for cations and anions. Remember, covalent compounds do not contain ions! If a covalent compound, use the prefixes in the compound's name to write the formula.

	Name of compound	Ionic compound or covalent compound?	Formula and charge of cation (if ionic compound)	Formula and charge of anion (if ionic compound)	Formula of compound
a.	sodium sulfide				
b.	sodium sulfate				
c.	sulfur dioxide				
d.	calcium hydroxide				
e.	potassium nitrate				
f.	barium chloride				
g.	phosphorus trichloride				
h.	sodium phosphate				
i.	calcium carbonate				
j.	calcium bicarbonate				
k.	nitrogen dioxide				
l.	potassium oxide				
m.	carbon monoxide				

2. Classify each compound as ionic or covalent—use either technique. Write the name of each compound. Remember that covalent compounds use prefixes as part of their name, ionic compounds do not use prefixes.

	Formula	Ionic or covalent?	If ionic, charge on anion?	Name
a.	NaI			
b.	$Mg(NO_3)_2$			
c.	CO_2			
d.	K_2S			
e.	K_2SO_4			
f.	SO_3			
g.	Na_2CO_3			
h.	$NaHCO_3$			
i.	P_2O_5			
j.	$Mg_3(PO_4)_2$			
k.	BaO			
l.	$Ba(OH)_2$			
m.	$LiBr$			
n.	CS_2			
o.	CaS			
p.	$CaSO_4$			

Exercise on Chemical Periodicity